Unleashing t…

An OrgZoo Quest Field Guide

Alex Bennet, Robert Turner, Arthur Shelley, Jane Turner

Illustrated by Mark Boyes

MQIPress
Frost, West Virginia
303 Mountain Quest Lane, Marlinton, WV 24954
United States of America
Telephone: 304-799-7267
eMail: alex@mountainquestinstitute.com
www.mountainquestinstitute.com
www.mountainquestinn.com
www.MQIPress.com

ISBN 978-1-949829-65-5
Cover drawing by MQI Resident Artist Cindy Taylor
Organization Zoo concept developed by Dr. Arthur Shelly
Original OrgZoo Characters drawn by John Szabo

Contents

QUEST XERCISES

Models

Learning Concepts 

Introduction

We at the Mountain Quest Institute join with Dr. Arthur Shelley from Australia and his beloved Organizational Zoo characters, along with illustrator Dr. Mark Boyes, on an amazing quest to unleash the human mind. Collectively, these OrgZoo animals represent *attributes and characteristics scattered throughout humanity*, with each animal representing a behavioral style recognizable in work and social settings, some perceived as largely positive and others perceived as largely negative, yet all potentially offering contributions to the diversity and discovery that illustrates the path of experiential learning. The guiding principles for using the OrgZoo characters in facilitation are in Resource A. Since these characters—affectionately referred to as "critters" in this text—are anthropomorphic or human-like in nature, their perspectives, reflections, and growth during this quest can be intensely insightful.

These critters are the voices in our heads, representing the behaviors with which we are both familiar and unfamiliar, comfortable and uncomfortable, as they whir around in our head. So, these adventurous critters are the ones bounding around in our (as yet untamed) mind, willing to get on with it and have fun questing. Some of them (such as the Kid, Dog, Gibbon and Yak, and maybe even the Mouse) surge forward without considering the potential risks. Nonetheless, as the Questers proceed and as they (and we) become more confident and competent, running off leash, the critters learn to choose the behaviors necessary to manage self in a given moment or situation. As these critters blend their unique cognitive abilities and feelings with their human attributes, learning to manage self, it is our hope that YOU can use this Quest Field Guide to experience *their* experiences, and choose which characteristics may be of service to you in unleashing the human mind.

This field guide is a companion piece to *Unleashing the Human Mind: A Consilience Approach to Managing Self*, which has at its foundation a ten-year Mountain Quest Institute research study exploring neuroscience and other scientific fields and disciplines related to experiential learning. The Intelligent Complex Adaptive Learning System (ICALS) theory and model resulting from this research is briefly provided in SECTION II of this field guide. Throughout this guide, the *Unleashing* in-depth discussion pages for various topics are provided in brackets following the appropriate text.

The learning quest the OrgZoo critters have selected is an expedition to an obscure remote region of Earth where they identified a challenging mountain peak reaching up to just over 12,000 feet. They have affectionately named their expedition and the summit Mountain Quest II. In preparation for the expedition, the OrgZoo critters initially convene at Mountain Quest in Frost, West Virginia, for an *Unleashing the Human Mind Workshop*. Mountain Quest resides at the foot of its own rolling hills, a portion of the Allegheny Mountains. After the workshop, the critters will participate in a group quest expedition to deepen their understanding and use of the new MQI Learning Theory and to prepare for their own individual and collective quest endeavors.

This OrgZoo Quest Field Guide will track the OrgZoo critters' ascent up Mountain Quest II. As the critters move through their quest, it also serves to introduce you to an *Unleashing* experience. For that purpose, this field guide provides shortened core ideas from the larger *Unleashing* text. You will find:

- New ideas are bolded.
- Key Learning Points are bolded and italicized.
- Quest Xercises for unleashing the human mind.
- Memory Nibs from the OrgZoo critters.

- Journal entries documenting the insights of OrgZoo Sherpas.
- Companion Resources.

Since our Questers are asked to keep a *Community Journal*, the bees volunteered to keep that journal, and they have made some of their thoughts available to you as *Memory Nibs*. Just as the point of an ink pen initiates the flow of ink, Memory Nibs initiate the flow of thought, jogging the remembering and integrating of what is already known with information coming in from the environment. The mind is an associative patterner; which means that incoming information in the present is associated with all the experiences and learning from the past and, in turn, provides the foundation for future decisions and actions. All participants were also asked to bring along their Self Phones for direct inner text. NOTE: Memory Nibs convey directly to the inner self. So, here's some of the things the OrgZoo critters are taking along on their quest:

An altitude tracker in the form of an altimeter is a part of their backpack so they can clearly see their elevation as they move up toward the summit. Each major mark is in 1,000-foot increments. There are special glasses for the critters to pull out and occasionally put on to ensure a fuller understanding of some new and complex learning concepts. For the critters, this takes the form of simple glasses—sort of mixed reality multifocals if you will—simple but special glasses to remind them to expand their vision, shifting their focal viewpoint to encompass a wider frame of reference, and intuitively recognizing the innate value of diverse ways of seeing and being. That's a lot for a simple animal to embrace, yet we believe that all of them who attempt this quest will make it to the summit!

We believe in you, too. As you use this new field guide, your understanding of your self will simultaneously deepen and expand. Follow these Questers throughout this field guide as they share some of their insights, and *invite you to participate in their journey*. As you track their quest, the critters will share their Xercises so that you can fully participate in their experience and—along with them—awaken new capacities in your own lives.

The Cast of Questers*

We're including those OrgZoo critters who joined the quest as well as those who chose not to join, and we have transcribed their thoughts on why they chose to participate or not participate in the *Unleashing* experience. Following those transcriptions, we include in brackets the characteristics attributed to each critter. The Beneficial Insect, Owl, and Quercus robur—renowned for their thinking skills as they navigate the uphill, icy slopes in their search for meaning and knowledge—have volunteered to serve as Sherpas. Sherpas are helpers, and heroes.

During the MQI *Unleashing* Workshop, the designated Sherpas are introduced to the role and activities of Sherpas. As part of their preparation, they will watch and study the thought-provoking 2015 documentary *Sherpa*, written and directed by Jennifer Peedom. Then they will watch a more recent documentary *14 Peaks: Nothing is Impossible*, written and directed by Torquil Jones. This 2021 documentary follows the unbelievable record-shattering expeditions of Sherpa Purja and his supporting Sherpas to the summits of 14 of the world's 8,000-meter mountains in seven months! The previous record for accomplishing this was seven years. Undoubtedly all of the animals find this awe-inspiring. So do we.

Leading this Quest

It makes sense that the Eagle has stepped up to lead this Quest. Eagles are inspirational leaders with fantastic clarity of vision and who can see for very long distances. They know what they want and soar well above the world, well aware of the potential opportunities and risks below and capable of reacting to them quickly and effectively. The Eagle brings a passion for achievement and success to this endeavor. He is speaking now to the larger group.

Eagle: "We have a wonderful group of Questers participating in the journey to the summit. Among many friends in the group, I note several Bees, who can be depended on to share what they are learning, and we have three Sherpas assisting us. First, the rare **Beneficial Insect** who is helpful, resourceful and forward thinking. Second, the wise and knowledgeable **Owl** who can assist with any issues that emerge. And then we are privileged to have the inspiring **Quercus robur** joining us on our Quest from whom all of us can learn and grow. When fitting, I invite all of our participants to take advantage of this great opportunity to interact with and learn from this amazing team of Sherpas." *[Intelligent, Visionary, Strong, Focused, Inspiring, Confident, Productive, Decisive, Balanced Sharp, Action oriented, Open Trustworthy, Intuitive, Aggressive, Accountable, Holistic, Extroverted, Respected]*

The Sherpas

Beneficial Insect: "This will be a great experience for all of us who are climbing together. I will try to help keep our focus on what lies ahead." *[Beneficial, Resourceful, Helpful, Positive, Forward thinking]*

Owl: "My role as one of your Sherpas on the Mountain Quest II expedition will be a life changing challenge for me. To help us achieve the summit, I will draw from what I know about you, what I learn at MQI about learning, and also about how Sherpas can serve you." *[Wise, Knowledgeable, Dedicated, Helpful, Intelligent, Logical, Respected, Sharp, Confident, Quiet, Communal]*

Quercus robur: "More than ever before, I will reach out to all of you across the biosphere of Mother Earth. You must know I will be with you through the lithosphere—land, atmosphere--air, and hydrosphere—water, wherever life is. While you see me as your connection to the plant world, remember that my massive oak tree roots are imbedded in the field of energy nurturing across creation. Go and I will follow where you go and I will be there with you in every step of your journey. Remember, when you call your Quercus I will hear you and we will ascend together. I promise you capacity and knowledge for your quest and celebration at the summit!" *[Inspiring, Enthusiastic, Knowledgeable, Decisive, Intelligent, Parental, Experienced, Stimulating, Reliable, Philanthropic, Approachable, Rare]*

Now, each of the Questers has the opportunity to share why they are joining this expedition, which will give you the opportunity to get to know them individually as well as in a group. The OrgZoo critters who have declined to participate also have the opportunity to share their own personal feelings about this quest. Following each of their thoughts, we include the characteristics attributed to each. We join them now.

OrgZoo Critters Participating in this Quest

Ants: "Right, we are off! What is the first task?" *[Busy, Hardworking, Communal, Loyal, Dedicated, Territorial, Present Focused, Instinctive]*

Bees: "Okay, guys, let's get started. We can do this together. Everybody do your part. We'll make it happen." *[Busy, Hardworking, Collaborative, Communicative, Communal, Sociable, Organized, Territorial, Intelligent, Knowledgeable, Loyal, Caring, Aggressive (when threatened)]*

Chameleon: "I believe this will help us all to blend into the environment." *[Two-faced, Cunning, Manipulative, Weak, Political, Intelligent]*

Dog: "Terrific, let's go! Which way do we start heading and how can I help?" *[Loyal, Trusting, Energetic, Enthusiastic, Boisterous, Gullible, Reliable, Predictable, Happy, Playful, Productive (when directed), Trustworthy (mainly)]*

Feline: "Who do I have as part of my team to ensure I can make this look good?" *[Individualistic, Agile, Aloof, Self-interested, Vain, Selfish, Frustrating, Arrogant]*

Gibbon: "This will be fun. I'm up for it and ready to do my part. Come on, I'll help you up if you get stressed." *[Happy, Playful, Energetic, Enthusiastic, Highly sociable, Fun, Interesting, Cool, Rebellious (if stressed)]*

Jackal: "Just don't get in my way. I'll be up front so I have the best view." *[Highly territorial, Zealous, Social, Aggressive, Agile, Controlling, Jealous, Fun loving (among their own group), Loyal (to their immediate clan), Caring (of those whom they consider worthy)]*

Kid: "Wow, do you all think I can make the summit? Does anyone want to place a bet?" *[Naïve, Playful, Energetic, Enthusiastic, Motivated (but without direction), Expendable, Youthful]*

Mouse: "I can't wait to get started on this quest and find out what I'm capable of achieving." *[Agile, Enthusiastic, Adaptable, Productive, Invisible, Busy, Individualistic, Intelligent, Economical, Reliable]*

Sloth: "Just be patient. [Yawn] I'll get started in a while." *[Slow, Weary, Minimalist, Submissive, Lazy]*

Triceratops: "I have to do this, but it's not going to be easy. Do we really know what we're getting into?" *[Weary, Pessimistic, Change-averse, Xenophobic, Battle-scarred, Tough, Negative, Slow, Reclusive, Reflective, Nostalgic, Problematical]*

Unicorn: "I am sure we can knock this over quickly and then take on another peak as well." *[(Self-rated) Perfect, Visionary, Open, Honest, Collaborative, Everything positive; (Rated by others} Arrogant, Aloof, Ambitious, Controlling, Dangerous, Frustrating, Insecure, Inconsistent, Nostalgic]*

X-Breed: "I have so much experience and talent, no doubt the team will rely on me all the way." *[(Cross-bred to perform} Arrogant, Verbose, Selfish, Ambitious, Beautiful, Busy, Fashionable, Extroverted, Educated, Capable]*

Yak: "You just wait and see. I'll be packing so much of the stuff you'll be glad I came along." *[Boisterous, Happy, Playful, Energetic, Enthusiastic, Tactical, Friendly, Social, Frustrating, Serendipitous, Focused (to the point of being unaware of potential collateral damage and risks)]*

OrgZoo Critters Declining Participation

Hyena: "I'll stay here and watch over the food supplies. [Heh, heh, heh!] You all don't know what you're getting into!" *[Aggressive, Scheming, Controlling, Manipulative, Strong, Political (Territorial), Communal, Opportunistic]*

Lion: "If OrgZoo is going to make it to the summit of MQII, you're going to need me to make it happen. Too bad I'm not going along." *[Strong, Powerful, Aggressive, Controlling, Lazy, Self-interested, Territorial, Manipulative, Confident]*

Nematode: "What? I have everything I need in my four inches of worm paradise, right here. Leave me alone." *[Lazy, Parasitic, Dependent, Self-centered, Invisible]*

Piranha: "No. Why would I need to work so hard at that when I have a smorgasbord here in my river?" *[Swarming, Aggressive, Ravenous, Dangerous, Selfish, Self-interested, Frustrating]*

Pestiferous insects: While found everywhere, did not make themselves available for an invitation. *[Arrogant, Ubiquitous, Ravenous, Self-interested, Swarming, Costly]*

Rattlesnake: "I've been up that way before, so I'll just hang out on a rock and warm myself while you're gone." *[Political, Sharp, Defensive, Noisy, Reactive, Insecure, Two-faced, Divisive, Survivor, Adapted to harsh environments]*

Vulture: "You all go on ahead. If any of you slip and fall, I'll be there to clean up." *[Nasty, Opportunistic, Scavenging, Self-centered, Dangerous, Political, Competitive, Sly, Patient]*

Whale: "While this Quest is outside the environment in which I can fully participate, I look forward to seeing the growth my earth-planted friends can achieve, perhaps reaching the same thought freedom in the air that we acquire through our beloved ocean waters. And I will be there to help when my specialized technical knowledge is needed." *[Intelligent, Inspirational, Knowledgeable, Powerful, Social, Communal, Self-limiting, Accident-prone, Reserved, Gentle, Endangered]*

*For more in-depth descriptions, see Shelley, A. (2007, 2021). *The Organizational Zoo: A Survival Guide to Workplace Behaviour*. Intelligent Answers.

Useful Definitions

It's always a good idea to develop a common definition of fundamental words and concepts used in any text since there are different opinions on even simple terms. So, in the context of the Quest Field Guide, we briefly describe the following terms: brain, mind, mind/brain, self, managing, consciousness, learning and knowledge, mountains, and Sherpas.

As will appear throughout this Guide, the **brain** consists of an atomic and molecular structure and the fluids that flow through the structure. The **mind** is the totality of the patterns created in the brain's 86 billion neurons and their firings across neurotransmitters and trillions of synaptic inter-connections. These patterns encompass all our thoughts. The term **mind/brain** refers to both the structure and the patterns emerging within the structure. [1]

Self is an emergent quality of living and developing as a human; the part of us that makes choices and carries out those choices through actions (if we choose!) And as part of a complex adaptive learning system, those choices affect every aspect of who we are as well as the relationships among those aspects. Self is in the decision-making role. In the uncertain, complex and continuously changing environment in which we live, managing self is of the utmost importance. In regards to self, we consider **managing** as an active and emerging role, not so much as controlling as choosing, that is, asserting our individuated agency in choosing the thoughts we think, the feelings we have, the beliefs we hold, and the actions we take. This means moving beyond the concept of total freedom connected to and perceived by an ego-focused "I" to a deeper understanding that your self is part of a larger Source to which you are always connected and with which you are continuously interacting. This is an underlying theme throughout the *Unleashing* experience.

Consciousness and self are terms that are often used interchangeably, and indeed we define self as the totality of the conscious and unconscious, the brain and the body. Digging down a bit deeper, consciousness requires awareness, knowing that we exist and others exist, which is the result of information coming in through our senses, but it is *more* than awareness. It is when this information is engaged in thought that consciousness occurs. Consciousness requires thought, the ability to link together internal and external knowledge such that it leads to a better understanding of our self and our environment, and the relationships among our self and others. In short, it is the sum total of who we are, what we believe, what we say, how we act and the things we do. As a working definition, we consider consciousness as a state of awareness and an understanding of the connections and relationships among that of which we are aware and our self.

Learning and Knowledge work together with learning being the creation of knowledge and with knowledge defined as *the capacity (potential or actual) ta take effective action*. This definition is compatible with the Western philosopher's description of knowledge as *justified true belief*. MQI draws from extensive national and international work with both learning and knowledge and has determined that careful consideration to both of these is fundamental to the future well-being of human existence.

Mountains are a particularly appropriate metaphor to use in a quest to expand the self and unleash the human brain. Mountains are more impactful on human experience than most of us would estimate. For example, mountains are said to serve as the water towers of Earth. Nearly half of our world's irrigation water flows from mountains. Likewise, mountains are the principal sources of fresh water on this planet, and even as the mountains offer refreshment to our bodies, they *exude freshness to the human spirit*. This happens in myriad ways. Individually and collectively, we are refreshed by their grandeur and their exclusive environments. Moreover, in many cultures and religions, humanity has been refreshed by the transcending discoveries of explorers, leaders and others throughout history who experienced epiphanies in mountain realms. Examples abound. Moses on Mt. Sinai; Mt. Fuji, near Tokyo, one of Japan's three sacred mountains, has been the destination of poets and pilgrimages for thousands of years; and Mt. Kailash in Tibet is sacred to four religions. The global list is rather endless. Indeed, mountain heights separate us from the terrestrial experiences of life and lift us to higher planes and viewpoints. They not only physically raise us up they also lift our gaze and elevate our thoughts. In their preeminence, we abound in perceptions and perspectives that enable us. The Quest to elevate our minds, to unleash the human brain, is as lofty as the mountains.

Mountain peaks like Mt. Everest require **Sherpas** to reach the summit. The famous Sherpas of Tibet actually are from the Sherpa people, a minority culture of Tibet which resides in an autonomous region of China and is the homeland of the exiled renown Dalai Lama. Sherpas revere their neighboring Mt. Everest, which rises 29,031 feet to the summit, and goes by the Tibetan name of *Qomolangma*, which translates as "Holy Mother". While many of the OrgZoo critters wanted to tackle Mt. Everest, it was wisely decided to first attempt a 12,000-foot quest to *unleash the human mind*.

Note that Resource C in the Companion Resources serves as a brief Glossary of Field Guide terms.

SECTION I: The Mountain—A Nexus of Choice

What is it to be human?
(Pre-Quest Workshop at MQI: Unleashing the Human Mind)

X1: Preparing for the Journey: Who is My "Self"?

Who we are, our *self*, ultimately comes down to the choices of what we *think, feel and do*. As we expand our understanding of self—and focus on how to manage self—let's start with a simple Xercise that explores the starting point of our Quest, that is, where we are, what our environment is, and how we feel about ourselves in this moment of life, the NOW. When repeated, this Xercise will help realize the beauty of the moment at hand.

STEP 1: As you move forward, you will want to experience this exercise several times during a single day, or perhaps once a week over a given time frame, such that you capture the variety of your life experience. If you have an alarm/alert on your watch or cell phone, you can set it to go off as a reminder. Then, when the alarm goes off, or when it is close to the set time, turn it off, stop whatever you are doing, and go to Step 2.

STEP 2: You are not in a hurry and you are in a quiet place. Stand up. With your eyes open, slowly turn completely around, scanning things that are close to you, say within five or six feet. Spend a moment thinking about each thing that comes into view. About each, ***reflect:*** How is this useful? Who created this? Does it serve a purpose? Do the pieces all fit together well? Are the colors pleasing? Allow yourself to be fully in the NOW; FEEL into the item upon which you are focusing, and allow each item their moment of importance.

STEP 3: Slowly, take a second complete turn, this time focus further out—into the distance, reflecting on all that comes into view. If you are outside, you may wish to take a third turn to reflect on the far distance. FEEL the energy around you and, again, give every item that comes into focus its special moment in time.

STEP 4: Close your eyes, and consider all of the things you have brought into your focus and *how they all fit together*. Now, imagine yourself as the model in a famous artist's painting, a *Carletti*, with all of the things around you of significance to the painting. What story do they tell? What is your relationship to these items? What do these things that surround you tell you about the figure in the middle of the painting, that is, YOU.

STEP 5: Taking a deep breath, leap into the future, say 100 years forward. Using your creative imagination, picture yourself in the middle of an art auction, and *there in front of you is the picture of YOU and all those things surrounding you*. Many people in the room are bidding, and the price gets higher and higher. As the people continue bidding, you look closely at the painting, and you begin to see *why* they are bidding. There is a relationship between the person in the painting that is YOU and all those things that surround you. They look REAL and yet there is an invisible energy that says more than the picture. You look closely at your face, captured with the eyes closed, and see a *wonder* expressed there. You note that the artist has captured the *knowing* that is occurring in the moment at hand. Even with the eyes closed, this figure has *full awareness of who they are and where they are*. The bids continue to get higher and higher.

STEP 6: Still in that place, you hear the auctioneer pause and say: "Is there another bid? This is a *Carletti*, a moment of beauty. The summation of the parts work together in such a way that nothing needs to be added, taken away or altered. The artist has captured a perfect life moment."

STEP 7: Still in that future place, you push your hand into your pocket and pull out a credit card with no limits, recognizing that in this place, at this future moment, you are wealthy and have the means to purchase this painting if you choose. *Do you choose to do so?*

REFLECT: What have you learned about your SELF from this Xercise?

Chapter 1: Being Human

Day 1 – Pre-Quest *Unleashing* Workshop at MQI
Notes and Activities

What does it mean to be human? We are infinitely complex beings with immense physical, mental, emotional and spiritual capacities. Presiding over our human systems, our human brains are fully integrated, biological, and extraordinary organs that are preeminent in the known Universe. Those are wonderful thoughts and words, only what does all *that* mean? Okay. Let's explore that a bit through a few of the areas introduced in *Unleashing*. Hmmm … What are the three main points …

First, **I am a complex adaptive system**. [2- 4] A complex system has a large number of interrelated parts which may or may not have nonlinear relationships, feedback loops and dynamic uncertainties difficult to understand and predict. They also have emergent properties which make the whole of the system (you) much more than the sum of its parts! For example, consciousness is emergent. An adaptive system—which can be responsive or proactive—improves its ability to survive and grow through internal adjustments. Complex adaptive systems operate at some level of perpetual disequilibrium, necessary for adapting but which contributes to unpredictable behavior. *Life is all about adapting*, and we will all agree that it certainly is complex!

One of the ways the human adapts is through **the plasticity of the brain**, which means you have a brain that is malleable or pliable. Plasticity is a result of the connections between neural patterns in the mind and the physical world, which means that what we think and believe actually impacts our physical bodies! [6-7] It's sort of a tit for tat kind of thing. The structure of our brain affects our thoughts, and our thoughts affect the structure of our brain. This is interesting to think about. Learning itself is dependent on modification of the brain's chemistry and architecture. We are always learning and changing. Forget what you learned in school, we are verbs, not nouns. And the bottom line is that the more you learn the more you CAN learn!

Second, **I have an inner and outer SELF**. [4-6] Self—an emergent quality of humans—is the totality of the conscious and unconscious mind, the brain, and the body, inclusive of everything it is to be human. From a quantum field perspective, the mind can be considered a field of potential—of possibilities—where at least half of the field has chosen to head the same direction. The mind is the seat of consciousness, enabling awareness of our self as a knower, an observer, and a learner, and as one who takes action, with our neurons forming a continuous memory of thoughts and actions, creating a coherent story of who we are.

Within each of us, embedded across cortical columns in the neocortex, is a personal ever-changing representation and models of the world [3]. The neocortex is the organ of intelligence, which receives input from movement, generates behaviors, and predicts the next input. Reference frames within the cortical columns enable the ability to perceive shape, changes, and locations relative to each other, providing the flexibility to process and navigate a changing, uncertain and complex environment.

Third, **I am on a learning journey**, what we can (hopefully) call an *Intelligent Social Change Journey* (ISCJ) [12-16]. This is a "self" developmental journey of the body, mind and heart, moving from the heaviness of cause-and-effect linear extrapolations, to the fluidity of co-evolving with our environment, to the lightness of breathing our thought and feelings into reality. These are phase changes, grounding in development of our mental faculties, each building on and expanding previous learning in our movement toward intelligent activity. [12-19]

We begin early in life with lower mental thinking based on logic: when you study you get a good grade on a test, when you break a rule, you get punished. [12-14] As we move into Phase 2 of the ISCJ, we begin to recognize patterns from past experiences that repeat themselves over and over. This recognition enables us to "see" (in the mind's eye) the relationships among events in terms of time and space. Thus, we are no longer in an action and reaction mode, but in a position to co-evolve with our environment in the present (the NOW), enabling us to better navigate a world full of diverse challenges and opportunities. Patterns grow into concepts (higher mental thought), and we begin to see larger connections among "things" as we search for a higher level of truth. To co-evolve with our environment requires us to develop a larger understanding of our self (values, beliefs, emotions, desires, etc.) and of others, developing empathy, which provides a direct understanding of another individual and a heightened awareness of the context of their lives and their desires and needs in the instant at hand.

The Chameleon

Emotions—present in all sentient complex adaptive systems—play a powerful role in this journey. Our emotional guidance system influences our perception of reality and how we respond to that perception [7-12]. They assign values to options and alternatives, often without our even knowing it! **Our emotions are a building block of consciousness.** While *emotional tags* are assigned to all incoming information, the entire body is involved in emotions. This was, and continues to be, essential to our survival, although largely today survival is not dependent on escaping a predatory animal! We're going to talk about emotions more later on.

There is a freedom that occurs as we leave behind the thinking patterns of Phase 2 of the ISCJ and open to the choices and discoveries of Phase 3. [14-16] As we enter this phase—co-creating the future—we are acquiring the ability to tap into the larger intuitional field that energetically connects all people, whether you call that an energy field, consciousness field, quantum field, or God field. This can only be accomplished when *energy is focused outward in service to the larger whole*, not constrained within, thus deepening our connection to others. Compassion deepens that connection. Thus, each phase of the ISCJ calls for an *increasing depth of connection to others*, moving from sympathy to empathy to compassion, as we become fully engaged in co-creating our reality.

Memory Nibs

- My thoughts change the structure of my brain!
- The more you learn, the more you CAN learn.
- We are verbs, not nouns like we were taught!
- Co-evolving requires understanding self and having empathy for others.
- My emotions are my personal guidance system!
- Wisdom is completeness and wholeness of thought.

The highest part of mental thought is **wisdom**. It represents the *completeness and wholeness of thought*. With the emergence of Knowledge Management in the late 90's, we initially saw a direct link from data to information to knowledge to wisdom, and that does somewhat describe mental growth in terms of developing increasingly higher order patterns. However, wisdom deals with both the cognitive and emotional, personal and social, as well as the moral and spiritual aspects of life. [19-21]

So, what does it mean to be human? We are superior energy beings in a physical body with a mind/brain, emotions, and a spiritual essence. We exist, we perceive our existence, with knowledge at our core providing the ability to think, choose, act, interact, learn, and change. We are complex adaptive systems, forever learning and changing, and cannot exist in stasis.

Quite simply, **learning is living; and living is learning**.

X2: Moving through the ISCJ

1

Cause and Effect

Reflect:

What early situations in life helped me understand the cause-and-effect relationship of my actions? How important were these lessons to my growth? Did these lessons need to be repeated for me to learn them?

2

Co-Evolving

Reflect:

Can you see the connections between past actions and where you are today? Do the patterns from the past help you make better decisions for the future? Do you appreciate the differences in people and their thoughts? How does this improve the world?

3

Co-Creating

Reflect:

Have you opened to the free flow of creative ideas? How do your ideas serve your Self? How do they serve humanity? Are there any negative consequences? Are you open to others' ideas? Are your beliefs or other mental models limiting your thought? Are you following your dreams?

Chapter 2: Environmental Currency

Day 2 – Pre-Quest *Unleashing* Workshop at MQI
Notes and Activities

Time accelerates. Distance shrinks. Networks expand. Information overwhelms. Interdependencies grow geometrically. Uncertainty dominates. Complexity boggles the mind. In short, we live in a **CUCA** world, that is, an environment of accelerating **C**hange, rising **U**ncertainty, and increasing **C**omplexity shadowed by the very human response of **A**nxiety. [23-24]

While change has always been around, today it's taking leaps and bounds in every area of life—*the pace of change has changed.* And there's so much uncertainty, even down to the point of survival. Complexity? Consider technology, medical care, water supply, power grids, the internet, politics, the economy, and the more complexity increases, the more unpredictable and vulnerable to failure, either from natural causes or intentional sabotage. All of which increases anxiety. And while it's nice to talk about all this as something separate, something being imposed *upon* humanity, with us in the role of victims, it is humanity that is the cause, and, to a large extent, the accelerated mental development that has accompanied our technological advancements.

Our inner and outer worlds are out of balance. [24-25] Accelerated mental growth accompanied by the sharp edge of win-lose competition has pushed individuals and organizations through ego into arrogance. While egotism says, "I am right", arrogance says, "I am right. You are wrong. And I'm not listening." *Egotism closes the door to learning, and arrogance builds forces that can lock that door.* But it's not too late! All this mental development does offer considerable opportunity for the future of humanity, that is, if we are able now through managing self to bring the mental into balance on the physical and emotional planes, with the spiritual counterbalance woven throughout, which means bringing truth, compassion, and unconditional love and purpose into our everyday lives.

Let's see, recognizing that we are setting the stage for Chapters 6-11 on managing self, what three key foci related to environmental currency should we like to address in this guide? Perhaps, learning in a CUCA environment, the impact of technology, and the potential loss of deep knowledge.

First, **my learning environment has a powerful influence on my thoughts and actions!** [26-28] In the world today, individual learning environments increasingly offer unprecedented richness of depth, reach, and variety. The world in its many dimensions may be summoned to present us in a few years, a few months, even in a few weeks what may have required a lifetime just a century ago. At the same time there is more offered that is false, misleading, or contrary to one's well-being or success than our predecessors on planet Earth could have imagined. Now, amidst all this variety and choice, our learning opportunities, inputs, and resources are often arrayed against our interests. Nevertheless, we have an immense number of choices available to cultivate our personal learning environment and serve our needs and interests. Also, we know that we have deep capacities in our brain to augment the personal learning environment that permeates our experience and helps manage our self. For example, mirror neurons, a form of cognitive mimicry, can transfer active behaviors and other cultural norms, often without our knowing this is occurring! This means that when we see something happening, our mind creates the same patterns of neurons that we would use to enact the same thing ourselves.

Cognitive mimicry transfers active behavior and other cultural norms [27-28]. This means that when we see something happening, our mind creates the same patterns of neurons that we would use to enact the same thing ourselves. Thus, the media and rhetoric we listen to, and the situations we put ourselves into, can strongly affect your thought, emotions and general well-being. Reflecting on this new understanding: How is the current world stage—filled with propaganda and violence—affecting learning and behaviors? How much of the repetitive negative activity in our systems is unconsciously being transferred to learners of all ages? What effect is political untruth having on our pandemic-honed population?

Second, **technology drives how I learn!** [29-32] In a social transformation beyond comprehension, exceeding the transformation from the Industrial Economy to the Knowledge Economy, we have become networked together at exponentially accelerating rates through technologies that compound their inherent capabilities and functionalities with each new iteration of innovation. As a result, the nature of learning has new unpredictable dynamics. As an example, in 2004 Sal Khan began helping a cousin with mathematics using an internet Yahoo app. [31] Before long, other cousins joined in his tutoring and he transferred his instruction to YouTube. Over the years, the Khan Academy (KA) platform has grown to approximately 8,000 videos that have had nearly two billion viewings. Now, 100 million users are calculated to have used KA worldwide in dozens of languages, and 10 million users subscribe to KA. On the KA website home page, we read "We're a nonprofit with the mission to provide a free, world-class education for anyone, anywhere."

In a world where the principal networking modality is the internet and there is increasing use of it, there is great concern regarding the quality and integrity of the flows. The current reality is that fake news, propaganda, bogus conspiracy theories, and misinformation in general are inordinately prevalent. Thus, learning in a networked world inherently needs increased and discerning learning skills for more reliable learning. Managing self requires self-managing our learning.

The creation of smart machines that are becoming autonomous—far beyond that which could be conceived a few years ago as we pursued artificial neural networks—is an offshoot of our accelerated intellect. [32-35] At the heart of everything we create in every field of human endeavor—health and medicine, safety, transportation, energy, exploration, manufacturing and technological innovation, finance, education—and in whatever work we pursue, *our brain/mind is augmented in increasingly unimaginable ways.* That augmentation is computers and advanced communication technologies, two technologies that are increasingly inseparable. Today brains learn continuously from interactions across the internet, globally networked crossing social, political and geographical boundaries. And interactive man-machine learning not only applies to how *we* learn from the technologies, but also *how the technologies learn from us*!

Memory Nibs

- The whole world is CUCA!!!!
- The pace of change has accelerated.
- Our inner and outer worlds are out of balance.
- Egotism closes the door to learning, and arrogance locks that door.
- Our brain is augmented by technology ... we are interdependent!

Third, **will this technological interdependency bring about the loss of deep knowledge?** [36-38] This is a question that bears reflection. Let's use a U.S. Navy model to understand levels of knowledge. *Surface knowledge* is predominantly but not exclusively information, answering the questions of what, when, where and who—and representing visible choices that are generally easily understood. In the form of information, it can be stored in books and computers and shared explicitly. *Shallow knowledge* is information plus some amount of context below the surface in terms of understanding meaning, and sense-making such that the knowledge maker can identify cohesion and integration of the information in a manner that makes sense. This meaning can be created via logic, analysis, observation, emotion, reflection, and even to some extend prediction. Shallow knowledge, which is *primarily social knowledge in nature*, emerging through interaction as people engage in conversations and dialogue.

Deep knowledge occurs when an individual has developed and integrated the following components in a specific domain of knowledge: understanding, meaning, insight, creativity, judgment, and the ability to anticipate the outcome of one's actions. It represents the ability to shift frames of reference as the context and situation shift, with the unconscious playing a large role. The source of deep knowledge lies in creativity, intuition, forecasting, experience, pattern recognition and the use of concepts and theories. It is the realm of the expert, whose unconscious has learned to detect patterns and evaluate their importance in anticipating the behavior of situations that are too complex for the conscious mind to understand. What is fascinating is that knowledge, which is the capacity to take effective action, is context sensitive and situation dependent! And while you can have deep knowledge in one field of knowledge, you might only have surface knowledge in another field!

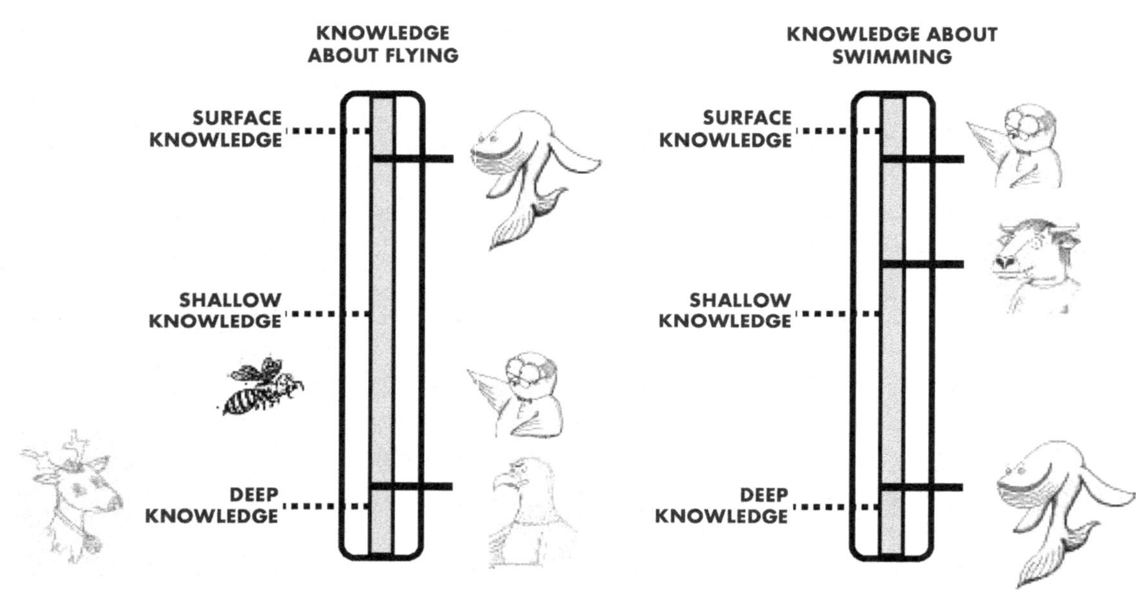

Let's ask our question again: *Will this technological interdependency bring about the loss of deep knowledge?* We might reflect that since technology makes MORE information available, and through social networks more opinions are expressed and ideas shared, that we have MORE knowledge. And this is certainly true! Note the differences between surface knowledge, which can be memorized and easily shared, and shallow knowledge, which is social knowledge, occurring through interactions with others, which enables conversations and the exchange of questions and answers, providing greater context. However, deep knowledge is largely experiential in nature, with the gaining of expertise through experiencing (acting) over time in a focused area of knowledge. Yet we also understand that as time passes and knowledge—which is always incomplete—expands, that deep knowledge does not necessarily remain "deep" but becomes more explainable as social knowledge (shallow knowledge), with newly discovered knowledge emerging through the depths of expertise. With further reflection, *how would you answer that question*?

X3: Using Your Self Phone

Bees: Communication skills are our specialty—a characteristic that is attributed to us in the OrgZoo text—and actually, you are equipped with this skillset as well, that is, when you choose to engage it. So, I want to spend a few minutes sharing how you might best use your Self Phone. This is the most important technology that you own! And yes, it is all yours. Only YOU can *manage your self*, and your Self Phone is a direct connection from your outer self to your inner self.

STEP 1: When you are clearing your head for the day, imagine your Self Phone in your hands. Relaxing into a comfortable spot, close your eyes, take a few deep breaths, and—watching the back of your eyelids—review the events, thoughts, and feelings of the day.

STEP 2: Ask yourself: What did you enjoy? What made your feel good? What are the points you want to embed in your memory?

STEP 3: Ask yourself: What would you change? How would you change it? And, in your creative imagination, make those changes. Now, reflect on this item again, only with the changes you've made in place. How do you feel about it now? What have you learned from this? Repeat this process as necessary.

STEP 4: Taking a deep breath, release any residue of negative thoughts or feelings and gift yourself a few minutes floating in the joy of being.

A NOTE FROM SHERPA PROF OWL:

Don't worry about how much stuff you purposefully send into your unconscious. There's *plentiy of room*. The internal knowledge dimensions of the inner self are far more vast than the external knowledge dimensions, no matter how many gigabytes your tablet or cloud account touts! And even as you age you are creating new neurons. That is what learning is all about!

There's a lot already on your "hard drive" of which you are not consicously aware, because your unconscious is processing continuously, taking note of the environment in which you place yourself even when you are not aware of it! And the really good news is that the unconscious mind, as part of your mind/brain/body system, exists to aid in your growth and survival. As such, it collects and stores information, experiences, feelings, and knowledge gained throughout your life. The beauty of this is that you have a "knowing" without "thinking" and all that is used to help you understand what actions to take in specific situations. That is your Self Phone getting your attention!

Chapter 3: Contemplating the Future

Day 3 – Pre-Quest *Unleashing* Workshop at MQI
Notes and Activities

A significant aspect of the mind/brain is its capability to continuously make sense of its environment and anticipate what's coming next. [38-39] We've all, both unconsciously and consciously, observed our mind's wonderful capacity to "predict". Unconsciously, this occurs with every step we take! On a more conscious level, when we first begin life, we very much live our lives based on past experiences. For example, as introduced as the first part of the ISCJ, early-on we begin learning that when we don't study for a test, we get a failing grade, or when we break a law, we get punished. As we recognize those patterns, we begin to move to the higher mental thinking of concepts—moving into Phase 2 of the ISCJ—and bringing together all three parts of time such that in the NOW we can look to past experiences (of ourselves AND others) and "predict" what we think will happen in the future.

This predictive capability is a ubiquitous function in the neocortex, what recent neuroscience research has called the organ of intelligence. [38] One way the brain anticipates the future is through the process of storing sequences of patterns. Since we never see the same world twice, the brain (as distinct from a computer) does not store exact replicas of past events or memories. Rather, it stores invariant representations. These forms represent the basic source of recognition and understanding of the broader patterns. In this way, you are able to apply memories applicable to a similar past experience to a current experience. We'll talk more about that in Chapter 6 when we talk about memory.

Thoughts are represented by patterns of neuronal firings, their synaptic connections and the strengths between the synaptic spaces. For example, a single thought could be represented in the brain by a network of a million neurons, with each neuron connecting anywhere from 1 to 10,000 other neurons! New incoming external information is mixed, or associated, with internal information, creating new neuronal patterns that represent understanding, meaning, and/or anticipation of the consequences of actions, in other words, *knowledge*. This *associative patterning* is a continuous process of learning. [40]

Contemplating the future is such a big topic! How do we possibly pick three main points from the *Unleashing* text? Let's talk a little bit about the importance of transparency, participation and collaboration, and then talk about the shift underway in terms of Singularity, and then forward a few thoughts about the future of work. Does that make sense? Three ideas.

First, **transparency, participation and collaboration are required to evolve in a networked environment!** [44-48] Transparency is being candid or open and free from guile. Participation is a keystone value for the Net Generation, who reach out worldwide to creatively engage people and ideas extending to both community service and political engagement, which reflects a *coherence between the heart and the mind.* Collaboration—which involves both engagement and participation—means working together. In a networked society, this includes reaching out to a fluid, ever-changing, interdependent network of diverse experts around the world.

Transparency, participation and collaboration are operational values that promote those two interrelated "t" words—*truth and trust*. Humans operate from a place of *yearning to know truth*. Truth is a living, dynamic awareness that expands our consciousness. Intelligence is defined as *an aptitude for grasping truths*. It is our neocortex—as the organ of intelligence—that is very much concerned with updating its models of the world based on a continuous stream of incoming sensory input, ever discovering higher truths, relating and connecting concepts to create new models of the world. Trust is based on integrity and consistency over time, saying what you mean, and following through on what you say. From a neuroscience perspective, trust in a relationship is very important for enhancing learning.

Second, **we are in the midst of a shift that will determine the future of humanity!** [52-55] A new generation with new ideas and experiences is coming to the fore, and the potential for large-scale change is afoot. Simultaneously, the breaking open of historical patterns and beliefs facilitated through the pandemic and life-and-death political and geographical power plays offer the opportunity for each and every human to consider their choices in terms of the future they wish to embrace.

The construction of a global culture, a culture of connection, and perhaps of coherence, is still a possibility. In fact, as boundaries are released between nations and connectivity continues to expand—given humanity can move beyond political and economic divisiveness—transparency plays an even larger role. With transparency comes a need for authenticity and global truth, greater freedom of thought for all, and the desire and opportunity for choice that leads to intelligent activity. A raising of awareness is coupled with knowledge resonance, that is, the free flow of ideas around the world, offering a bouquet of choices for each person. And coupled with the openness to new ideas and learning is *thought attendance*, that is, a rising awareness that form follows thought (we talk about that in Chapter 5), a consciousness of our own consciousness, and the thoughts and feelings emerging from self. With that awareness empathy emerges, a stronger, more intimate form of feeling and connection with the other, while in the larger populations a resonance occurs, with new ideas precognating the coming renaissance, a new Golden Age of Humanity.

Is that the *point of Singularity*—a hypothetical point in time when artificial intelligence (AI) outdistances human intelligence—that so much of our technological growth has prognosticated? [55-59] We forward that the Singularity will most likely occur in concert with a human conscious shift—which is underway now—as part of our human journey, the instant where WE surpass our current level of thinking, whether expressed through technology or through the expansion of our consciousness and the mastering of our mind/brain. Only, perhaps our interpretation of what is happening may play out differently than expected!

Consider the possibility that at the same instant where technology catches up with the mind/brain operating at the historic level of consciousness that the human mind/brain takes a creative leap, a time when humans have advanced technology so far and fast that it enables humans to transcend their biological limits. Reflect that the way we define "Singularity" echoes our current limitations; when those limitations are removed and the mind/brain is unleashed the term "Singularity" represents a higher level of technological advancement. Thinking from the quantum field perspective, we recognize that behind a closed door "all that is possible exists", and that prior to opening the door, what is behind that door is potential. Once that door is open, our frame of reference has shifted such that what "is possible" AFTER the door is opened can be quite different than what "is possible" BEFORE the door is opened. [58-59]

Third, **the future of work is changing!** [60-63] There is no question that technological advancements—particularly AI, machine learning, and human augmentation, and the potential of developing artificial general intelligence (AGI), which would more fully exhibit human-line intelligence—are transformational at all levels of the human experience. This is transforming the way we think about reality and the role that we play as humans in that reality! That said, it is important to recognize that while technology continues to have momentous impacts on our lies, the unfolding of robotics, AI, and AGI is an evolving, gradual process. Even spectacular advances reflect substantial time lags from discovery to wide-spread adoption.

This does not mean that AI and automation won't continue to impact the *types* of jobs and skills needed by humans to move into the future, which are very much dependent on economic incentives as well as policies and decisions put forward by public and private organizations. And of course, human choices are necessary to direct the outcomes of technological development. Further, a reliance on human-generated data produces biases, which adds "noise" to the system that can only be recognized and mitigated by human intervention. In other words, humans are still needed in this process!

Acknowledging that humans are complex adaptive systems, with the mind/brain designed to adapt to new emerging situations—a depth of ability and flexibility which to date has not been replicated through AI—humans will remain necessary in decisions processes, with the use of judgment very much involved with emotions. Humans also excel at a variety of other tasks which involve social interaction and unpredictable physical skills as well as the use of commons sense and general intelligence. In looking at the future, specific skill sets—and success factors—emerge throughout this text. Assuming our Sherpa Prof Owl will lend us his chalk board for a few minutes, we'll see if we can briefly connect some of the dots.

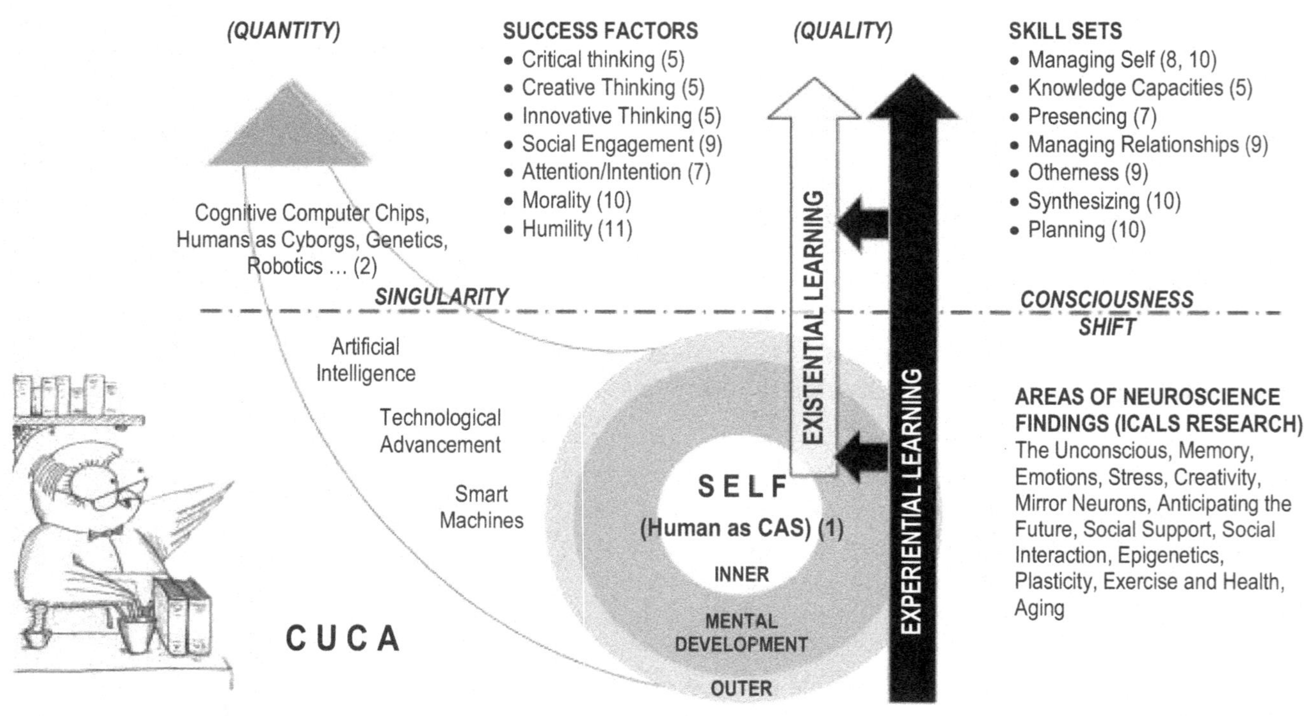

Success factors and skill sets for unleashing the human mind as we move into a new Golden Age.

Self is at the core of this graphic, and as our technological advancement continues to occur, the graphic takes into account the Singularity/Conscious Shift that is currently underway. The specific skill sets spelled out to the upper right are those necessary to navigate the changing nature of work. At the very top is managing self, which is interwoven throughout this text and addressed from the viewpoint of the body, mind, emotions and soul as we progress developing field guide notes from Chapter 6 through Chapter 11. The number in parentheses following the listing of skill sets and success factors is the *Unleashing* chapter where these skill sets appear. They are also all mentioned in those chapter segments in this field guide. Experiential learning is addressed in Chapter 4, and existential learning is introduced in Chapter 10. The areas of neuroscience findings at the lower right—which support development of the ICALS model—are embedded throughout the text of *Unleashing* and this field guide. Resource A is a listing of the neuroscience findings foundational to the ICALS model. So, throughout the rest of this book you will notice all of these subjects emerging as we slowly but consistently begin to unleash our minds, recognizing the power of those minds, and exploring the world around us through different viewpoints and paradigms. Welcome to the future.

SECTION II: The Basecamp—Adapting for Change

Finally, they are here, at basecamp 5,000 feet up on the mountain! Admittedly, the critters hopped a ride on several Range Rovers most of the way. Still, there was a sense of accomplishment. While the Pre-Quest Unleashing Workshop held at Mountain Quest Institute—which was itself situated in the Allegheny Mountains of West Virginia—had been, well, interesting, it just wasn't the same as actually being up on the mountain! Everyone was needed to help set up basecamp, where they would stay for two days to become acclimated to the thinner air.

There is something remarkable about being in an elevated rarefied atmosphere isolated from your normal world and at the very edge of your capabilities and comfort zone. You know you have committed to the journey now, and that it is too late to back out, but the small voices of doubt start nagging in your mind (especially for Triceratops, and if they had come along, it would be the same for the vulture and piranha). That little voice in your head asks: What if I cannot do this? We are doomed to fail! Why is this important—is it worth it? At this stage your adaptability is being tested. Can you balance these voices with positive aspects to enable you to make the creative leap out of your comfort zone?

Chapter 4: A New Theory of Learning

Day 4 – Base Camp Day 1 (Altitude Adjustment)
Notes and Activities

When we talk about learning, it's not uncommon to pull up a vision of the brain, which is the molecular structure and the fluids that flow within and through that structure. But it is the mind, created by neurons and their firings and connections, that is the totality of the patterns in the brain and throughout the body, which encompasses all our thoughts, ideas and perceptions. [65-66]

As we interact with life our neuronal circuitry is continuously changing. This can happen because neurons are not physically bound to each other so they have the flexibility to repeatedly create, break, and recreate relationships with other neurons. This is the process of plasticity. And this is continuously happening between the *billions of neuron cells and the trillions of synaptic connections among them* that reside in YOUR head!

As all those neurons in your head are continuously changing, YOU are learning, and this is the creation of knowledge as you go about experiencing in your everyday life! Today, because of the technological advancements that enabled brain measurement instrumentation such as functional magnetic resonance imaging (fMRI), we can measure what is happening in our brains. Researchers are now even able to explore the unconscious mental life, expanding our knowledge of learning from the inside out. Otherwise, this Quest Field Guide and *Unleashing the Human Mind* would not have come into existence.

There is so much we understand now. And while the more you learn the more you realize there is to learn, still, understanding that **we are complex adaptive systems** is a big step, and glimpsing from the inside-out at how our mind/brain works adds a new level of understanding to the concept. [Considering the magnitude of neurons and their connections in the mind, there is no question regarding complexity, and the amazing idea of plasticity (with neurons ever changing their connections) certainly shows adaptation, which improves our ability to survive and grow through internal adjustments. And the final piece needed to describe our learning process is intelligence, the capacity for reasoning and understanding or an aptitude for grasping truths. Thus, we can describe ourselves as an Intelligent Complex Adaptive Learning System (ICALS), which is the appropriate name of the new updated experiential learning theory.

There are important aspects to understand about the ICALS, which is YOU. So, let's talk about self-organized learning, the five modes of learning, and the guidelines for learning from the ICALS research. That will help you understand your learning processes at a new level.

First, **my intelligent learning system is a self-organizing complex adaptive system**. [70-71] That means we have a great deal of freedom for organizing, developing, and expanding our self so that we can achieve our desires and objectives. All this is happening inside our brains and minds and, well, actually, our whole body because the brain and the body are tightly connected through the flow of chemicals such as hormones and neuropeptides. Everything is entangled. Even neuronal signals flow through the central nervous system and we all know that the flow of blood continuously provides energy to the brain and body.

The environment is also closely connected to our mind/brain/body via incoming information and exchanges of energy with the body and from physical interactions between the body and its environment. If we take a holistic view of all this, we can realize there are *multiple* entangled complex adaptive subsystems which co-evolve through various levels of sensing, information and energy transfer, and some level of mutual structural adaptation. There's that word again; it's clear *choosing and adapting* are essential to living! And while from our individual viewpoint, the conscious self is one unifying perspective of all this activity—and the self provides the foundation for all our learning—our conscious self does not control all the subsystems of the mind/brain/body. Feelings, memories, belief systems, past knowledge, motivation, and immediate goals all influence learning effectiveness, with much of this embedded in our unconscious.

It is probably clear now—and you know this yourself from your own learning experiences—that such a system can rarely be "controlled" by force. Rather, it must be guided and nurtured through supportive influence. And the beliefs and attitudes of the learner affect learning and *can be under the control of the learner if he or she so chooses.* One of the things we understand from our research is that voluntary learning promotes theta waves which correlate with little or no stress and positive feedback, which fosters a higher level of learning. [84] Theta waves are low frequency (4-8 Hertz) electrical waves generated within the brain. Thus, your "will"—your desire to learn—plays a VERY large role in your capacity to learn!

Second, **while there are five modes of learning, I have my preferences!** [74-78] Building on the work of John Dewey, Kurt and Jean Piaget, in the late 1980's David Kolb published a model of experiential learning that combined cognitive theories (which heavily emphasize abstract symbols and their manipulations) and behavioral theories (which ignore consciousness and subjective experiences). In 2002 biologist J. E. Zull expanded the Kolb model, suggesting that there was a related view of the human brain and that Kolb's learning cycle arose naturally from the structure of the brain. In 2010 nuclear physicist David Bennet completed a ten-year study of emerging neuroscience research focused on experiential learning. Looking through the lens of consilience, thirteen areas critical to adult experiential learning were identified in this research (see Resource C). Through this process a fifth mode of social engagement was added to the original four modes, and the significance of the role of self as an underlying foundation and the influence of the environment in the learning process was acknowledged.

FIVE MODES IN THE ICALS MODEL

Mode and theoretical descriptor	*Characteristics when this mode is engaged*
Concrete Experience (grasping through apprehension)	= Sensing, feeling, awareness, attention, intuition
Reflective Observation (transformation via intention)	= Understanding, meaning, truth and how things happen, intuition, integrate, look for unity
Abstract Conceptualization (grasping via comprehension)	= Concepts, ideas, logic; problem solving; creativity; build models and theories; anticipation of outcomes; control, rigor, discipline
Active Experimentation (transformation via extension)	= Act on environment, focus attention, object-based logic, heightened boundary perception, sensory feedback to brain
Social Engagement includes Social Support (grasping through direct comprehension) and Social Interaction (transformation via association)	= [Social support] open mind, risk-taking, willingness to listen and learn, reducing stress and fear, creating resonance with people and ideas, contributing to evolution and sculpting of brain [Social interaction] accelerating learning and creativity; enhancing understanding, meaning, truth and how things work; developing a shared language; supporting use and understanding of concepts, metaphors, anecdotes and stories

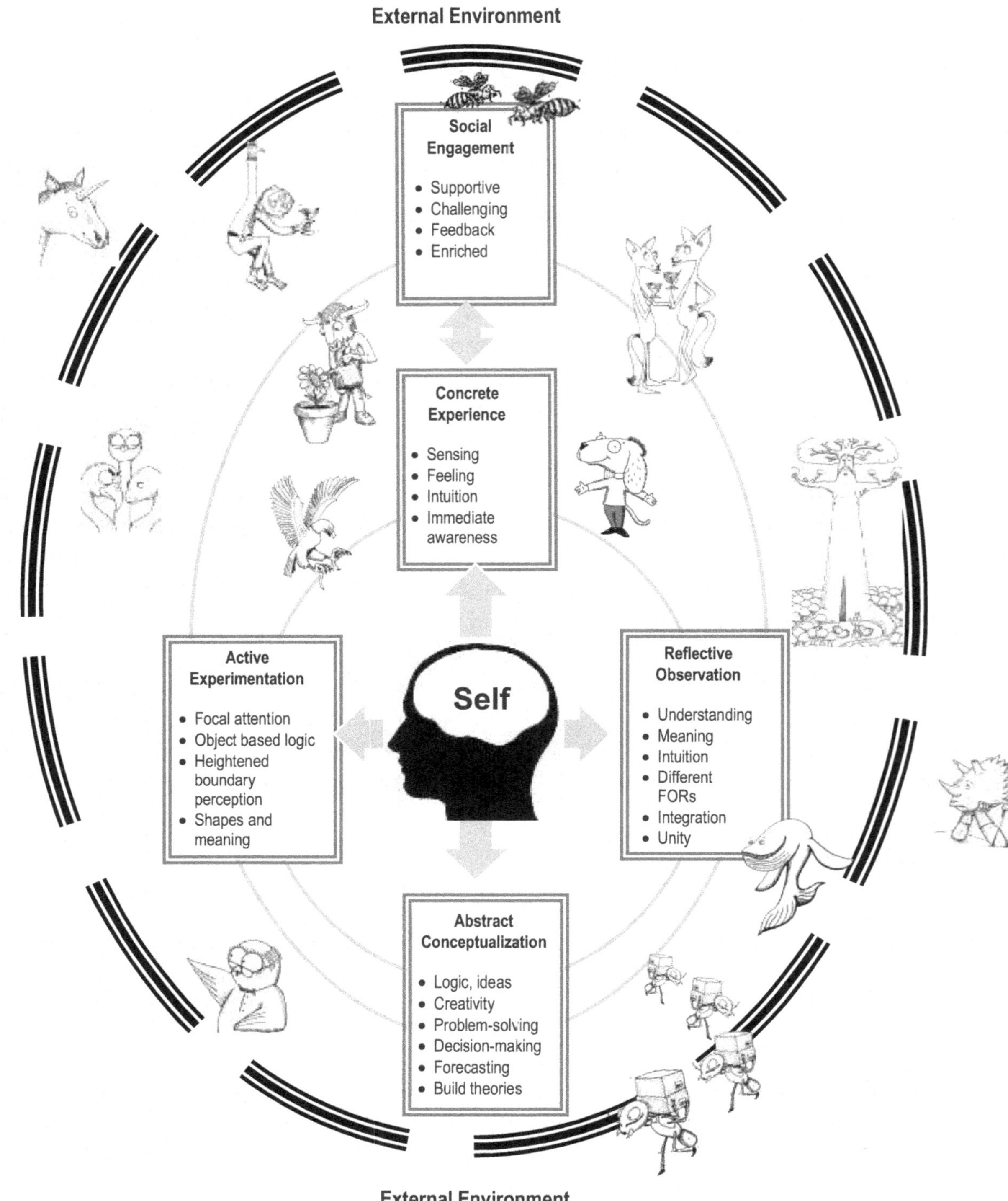

The Intelligent Complex Adaptive Learning System model for experiential learning with some of the Questers choosing their preferred mode of learning.

Third, **guidelines for learning emerge from this research that will prove useful to me**. [319-325] These guidelines offer a high-level perspective for learners as you acquire and apply knowledge and consider reprioritizing the development of your own self. Since everybody is a bit wiggly after that rather deep stuff Sherpa Prof Owl was spouting—and since our Questers have looked ahead and read these—let's ask them to share in their own words what is useful to them.

Kid: "Let me start! Let me start! Can you believe that MY *mind has no limits*! All I have to do is eat well and exercise, and playing sports does that! And maybe I need to learn to focus a bit; it's kind of hard to remember to do that. I'm going to pay more attention to you guys and maybe listen a bit more on this expedition. There is just such wonderful stuff going on all around me!"

Triceratops: "I have something to say. You know, I remember pretty much everything that has happened in the past, and unlike this kid, I don't see this world as such a great place. I've seen some pretty bad stuff; and bad stuff is happening pretty much everywhere today. Now, usually I don't like to say how I feel to anyone, so my speaking right up front is unusual. But I've been reflecting on the role our beliefs play, and especially what was said about *when a person believes they cannot learn then they won't be able to learn*. And I'm going to reflect more on that as I head up the mountain."

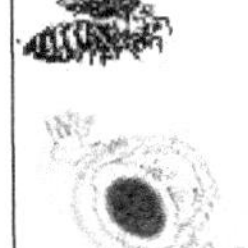

Bees: "We're next. It's very important to recognize the *influence of the environment*. That reflects what we think and the way we feel, whether we are aware of it or not! It will be hard to not be aware of that this next week!" **Gibbon** (jumping in): "Yes, yes. That's what I was going to say. I mean; besides our relatives, we can choose our environment a lot of the time like in the movies we watch and who we talk to and what trees we swing from. So, this is really important!"

Sherpa Beneficial Insect: "While there's a pause, I'm going to interject in here that I've discovered that *with knowledge comes responsibility*. There's some wisdom with this. When we learn and create knowledge, we have a duty to use that knowledge responsibly. As a Sherpa, I take this very seriously, and I think we all should. Thank you for listening." [Nods]

Chameleon: "You know something I never thought about is that my unconscious causes a lot of my thoughts and feelings and actions. So, a lot of how I act is not my conscious choice. [Skeptical looks] No, really, it's like just habit, even though I think it's good. My unconscious is working 24/7 just for me! Liked hearing that *my unconscious never lies*, but then it kind of upset me that it's not always right, that *it can make mistakes*! I put lots of good stuff into it!"

Yak: "I strive for certainty. Doesn't everyone? So, this discussion about risk was new to me, although the good news seems to be that there's no risk aversion, there is only risk management, sort of like managing self. Being on this mountain is a good time to think about this. I read that risk requires learning, and learning creates knowledge, and that leads to action, and actions need wisdom, with all these things connected in circular spirals that actually determine the quality of our lives, and even our existence!"

Dog: "There's the dinner triangle! Only let me say one thing here as we're facing this climb together. I was pleased to read about the *sacredness of values*, the importance of dedication and devotion whether to a purpose or a person, and that values are inherent in all learning." **Jackal** (interjecting loudly): "Especially to your family." **Quercus robur** (softly, but heard by all): "To everyone!"

Chapter 5: Thoughts and Thinking [85-111]

Day 5 – Base Camp Day 2 (Altitude Adjustment)
Notes and Activities

The human ability to control our own attention, remember, think abstractly, and reason are largely what sets us apart from other animals. You OrgZoo critters must be super aware of this. With these abilities humans have built cultural systems such as language, religion, art, and science. Whether at the individual, organization, country or global level, humans require ordered systems with rules to engage in intelligent activity, which at the highest level is called civilization.

It takes the mind and development of our mental faculties to create knowledge, that is, to use our learning to *effectively act on and in the world.* We are mentally focused in the physical, with neurons firing not only in our minds, but throughout our body, with connective tissue to all parts of our body. Everything about mental work can be applied to the physical world. And there is so much we are learning about the power of our thought! Here are a few highlights that might just shift your thinking.

Energy follows thought. [87-88] Everything in our material reality was a thought first. Therefore, the physical world is an "effect" not a "cause". Your thoughts count! A related ICALS finding is: *Thoughts change the structure of the brain, and brain structure influences the creation of new thoughts* [88]. This recognizes the interdependence of the mind and the brain, with the accelerated mental development associated with technological advancement occurring now producing a higher level of mental thought, which in turn changes the structure of the brain to enable an even higher level of mental thought, and so forth. As you can see, the learning you engage in today makes a big difference to your future capacity to learn.

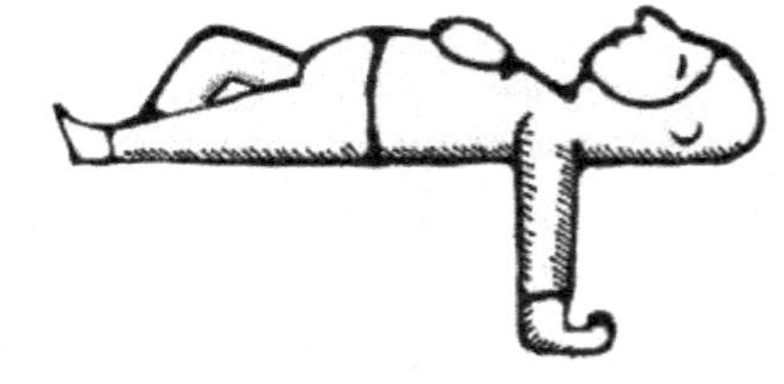

Fortunately for our OrgZoo Sloth, *thinking is mostly unconscious* [86]. Thought is a combination of choices, connecting a choice to other choices previously made, and affecting further choices by limiting or reducing those choices, in all of which the unconscious plays a primary role.

Okay. Three core ideas. We need to say more about conceptual thinking, so how about: higher mental thought (concepts) builds on lower mental thought (logic), critical thinking helps us make sense of the world, and every person is creative. And, along the way—when talking about critical thinking—lets introduce Knowledge Capacities, and we have a particularly applicable one for this field guide!

First, **higher mental thought builds on lower mental thought**. [89-91] This idea was introduced in Chapter 1. When we are able to discover patterns in life based on the lower mental thought of logic, we have moved into the higher mental thought of conceptual thinking. From this viewpoint, we are looking for logical examples that demonstrate the concepts and the relationships among concepts that are emerging in the mind. As we discover examples that lie outside of the concept, the concept is no longer true and must be shifted to include these new examples, ever asking *Why?* and expanding our level of truth. The more examples that are brought to bear, the greater the potential to find a higher level of truth. As these examples are shared in a cooperative and collaborative fashion, ever open to a diversity of insight and reflection, there is a balancing of thought and perceptions that occurs and an expansion of consciousness.

It is more difficult to recognize our mental senses than our physical senses. For example, when you look at another individual what do you see? A first response might be descriptive characteristics of the individual and the backdrop against which you are seeing them. But we now know that we sense through three planes simultaneously, and that any emotional tags connected with that person or situation will be connected to what is visual and what is "felt". Because mental thoughts are not only past memories and learning but very much connected to the NOW (how you feel about what you are experiencing), and of an emergent quality, and because the conscious mind is of a linear nature, information being processed by our mental plane is harder to separate out. Yet, it is very much there. *In this journey of discovery, we are using our mental faculties to co-evolve with—and co-create—the world within which we interact.*

Second, **critical thinking helps us make sense of the world**. [91-98] In our search for larger concepts, larger truths, reconstruction, assessment, and evaluation are used to discern the information; to determine its status as deductive, inductive, or irrational; and to judge its accuracy and reliability. Deductive reasoning is a logical progression, with the conclusion based on multiple premises assumed to be true. Inductive reasoning starts with an observation or question, and then, by examining related issues, tries to figure out a theory. Both are involved with the search for truth.

The intent of critical thinking is to allow individuals to form their own opinions of information resources based on an open-minded evaluation of the features and components of those resources by using, for example, problem-solving skills. These skills would include listening, considering other points of view, negotiating, and evaluating one's own mental models—which are patterns of neuronal firings—including beliefs, values, and perceptions. Critical thinkers consider not only the data and information, but the context of the resource, questioning the clarity and strength of reasoning behind the resource, identifying assumptions and values, recognizing points of view and attitudes, and evaluating conclusions and actions.

Knowledge Capacities [98-101] provide focused thinking approaches in support of critical thinking, different ways to perceive and operate in the world around us, that change our reference points. In a CUCA environment where surprises emerge and must be quickly handled, capacity is more important than capability for sustainability over time. For example, the building of capacity occurs when you observe the characteristics and behaviors of the OrgZoo characters in your own personal organization zoo and build self-awareness of your own characteristics and behaviors, which we call *instinctual harnessing*. Because it provides a removed perspective, it is possible to dissociate from personal feelings and more objectively (and metaphorically) view a team, organization, or the world as a collection of creatures interacting in, and with, their environment. This helps you understand moods, inconsistencies and vagaries, know others strengths and weaknesses, and learn how to manage through these.

Let's play with an Xercise called Focused Trait Transference, a short process that provides a focal point for looking at specific animal traits that correlate to human traits, helping us to understand ourselves and others. Remember, the human is a complex adaptive system with incredible diversity and individuation. In this Xercise seven diverse animals are identified with the hopes that each individual will be able to identify with at least one attribute from each animal. Then, understanding the set as a whole and how they work together begins to honor the complexity of the human, that is, YOUR COMPLEXITY!

With the advent of the internet, and the increase in the amount of data and information available, critical thinking becomes essential to functioning in the world. The successful self, a critical thinker has a colossal collection of different approaches to deal with different situations and contexts. This includes meta-thinking, or *thinking about our own thinking and decision processes* in order to continually improve them.

X4: Focused Trait Transference

STEP 1: Make a copy of the list below of the seven different animals with short descriptions of traits that they represent. Read the descriptions for each animal. Because these are animals most likely familiar to you, perhaps you are aware of additional traits. Add these to the list.

STEP 2: Reflect carefully on the traits for each animal. *Ask*: Are these traits that I see in the people around me? Are any of these traits that I have or exhibit to others? Highlight the traits for each animal that you see in yourself.

STEP 3: Bring all the traits that you see in yourself together as a set. Identify the strength of each trait in terms of their value to you. *Ask*: Which ones are the strongest? How do these traits work together? How have these traits been valuable in the past? How do these traits impact my interactions with others? How do these traits impact my work? How can I use these traits to my benefit and the benefit of others with whom I interact? *or* How can I change these traits to add more benefit to myself and others?

STEP 4: Consider the value of this exercise. *Ask*: What have I learned about myself and my interactions with others?

INCLUDE JACKALS

At least they had the good sense to include CATS!!!

THE LIST*

(1) ***Elephants.*** Symbolically, elephants represent *strength and power*, especially power of the libido. Their strong sense of smell represents higher forms of discrimination leading to *wisdom*. Elephants also show *great affection and loyalty* to each other and their families. In their behaviors are the ideals of true societies.

(2) ***Mountain gorillas***. Mountain gorillas, *strong and powerful*, are generally *gentle and shy*. They are *highly social* and live in relatively *stable, cohesive groups*, with long-term bonds developing between the males and females. The dominant males mitigates conflicts within the group and protects the group from external threats.

(3) ***House cats.*** The cat represents the attribute of *independence* and a wide variety of traits such as *curiosity, cleverness,* unpredictability, unsociability and *healing*. Because they can see in the dark, they are often associated with mystery and magic. Shelley (2007) attributes the following to cats: individualist, agile, aloof, self-interested, vain, selfish, frustrating, and arrogant.

(4) ***Songbirds.*** This group of perching birds (*Passeriformes*) includes over 4,000 species of birds found around the world, all equipped with vocal organs that produce a diverse and elaborate bird song. Both physical and spiritual, sound is an expression of energy. For example, the canary reflects the *awakening* and stimulation of the throat and heart, which gives increased *ability to feel and to express feelings*. Since the power of voice, music and sound is within you, what you say and the way you say it will be keenly felt by others.

(5) ***Horses.*** Key concepts connected to the horse are travel, *freedom* and *power*. The horse's energy is expansive, historically serving people in agriculture, recreation, war and travel, enabling people to explore the world and discover freedom from historical constraints. They signify the *power* and *movement* contributed to the rise of civilization, and have been poetically connected with the wind and foam of the sea.

(6) ***Dogs.*** The two words that immediately come to mind for dogs are *faithfulness* and *protection*. Other terms associated with the dog include companionship, nurturing and caring and guardianship. Shelley says that dogs, a highly versatile and enthusiastic group of creatures, are loyal followers, not leaders, with the behavior highly dependent upon their master (leader) and their environment. He attributes the following to dogs: *loyal, trusting*, energetic, *enthusiastic*, boisterous, gullible, reliable, predictable, *happy, playful*, protective (when directed), and trustworthy (mainly).

(7) ***Rabbits.*** Connected to fertility and new life and imbued with *ambition, finesse and virtue*, the rabbit brings with it the *sensitive* and *artistic* powers of the moon. Moving in hops and leaps, it is fleet of foot and active both day and night. Although associated with fear by some, the rabbit has wonderful defenses, clever at doubling back and making quick and rapid turns.

*These characteristics and short descriptions are pulled from T. Andrew's *Animal Speak: The Spiritual & Magical Powers of Creatures Great & Small* (2005) as well as A. Shelley's *Organization Zoo: A Survival Guide to Workplace Behavior* (2007, 2021).

Third, **creativity is an attribute of every healthy living human**! [101-109] Creativity is the emergence of new or original patterns such as ideas, concepts, or actions. It is the ability to perceive new relationships and new possibilities, seeing thing from a different frame of reference, realizing new ways of understanding, or having insights. What a lot of words! But you get the idea. And now we know from neuroscience—and from our own common sense—that everyone is creative, and thinking more creatively is a choice, with creativity an inherent capability of every mind/brain. The commonsense part of recognizing that becomes obvious when looking at the growth of a child, or your growth. As a youngster, everything was new and had to be considered and learned from and, well, acted upon. That took a lot of creativity! Learning is creative! Learning consists of building new patterns within the mind that represent ideas, concepts, and capabilities, and processes that are internally imagined (creative imagination), or represent some interpretation of external reality.

There are lots of ways this ordinary creativity occurs, through free-flow and randomly mixing patterns or the accidental association of patterns, both of which create new patterns. While the unconscious plays a big role, both conscious and unconscious patterns are involved with creativity. In fact, *volleying between the conscious and the unconscious increases creativity*, with the unconscious potentially producing *flashes of insight!* How does this occur? This shifting between conscious and unconscious thinking makes use of the memories and knowledge in the unconscious as well as the goals and thinking of the conscious mind. This increases the chances of associating conscious ideas to create new ones, which supports innovation. While this occurs in the transition from the waking state to the sleeping state (and vice versa), to do this while awake requires some level of control and discipline in implementation. The body asleep/mind awake approach achieved through hemispheric synchronization is an example, which we describe later.

Becoming aware and understanding that *every healthy mind is capable of creativity*, and that the rules and practices of creative thinking are available to anyone, opens the door to enhanced individual and organizational creativity and expanded learning capacity. And, remember, the unconscious plays a LARGE role in creativity, and since your unconscious works 24/7 for you, that's a lot of creativity!

When creativity has a utility, leading to a new product or process, that is innovation. **Innovation** means the creation of new ideas *and* the transformation of those ideas into useful applications; thus, the combination of creativity and contribution as operational values promote innovation. [109-111]

As an idea generator, knowledge is the currency of creativity and innovation, and knowledge cannot exist without information. As information flows freely and is generally available to all—and as people recognize the power of and creative potential in the mind/brain and learn to tap that power and creativity to create change for the greater good—each individual has the ability to be an innovator, that is, providing a "new" product or processes in the domains of their life!

Past experiences, feelings, knowledge, goals, and the situation at hand all influence how creative an individual will, or can, be. It is the context of the activity or situation at hand (need, challenge) that triggers the putting things together (bisociation) in an unusual way to create (and recognize) something that may be new and potentially useful (innovation).

Memory Nibs

- Energy follows thought!
- My thoughts actually change the structure of my brain!
- And that affects my thoughts! Wow!
- Thinking is mostly unconscious.
- The mind is always seeking higher truths!
- Critical thinking support effective thinking.
- Everyone is creative, even me!

Eagle: Several of our Questers have a question related to the types of thinking that were presented in today's learning. Basically, how do these things relate? Specifically, how do conceptual thinking, critical thinking, Knowledge Capacities, creative thinking, and innovative thinking relate? Are they different classes of thinking?

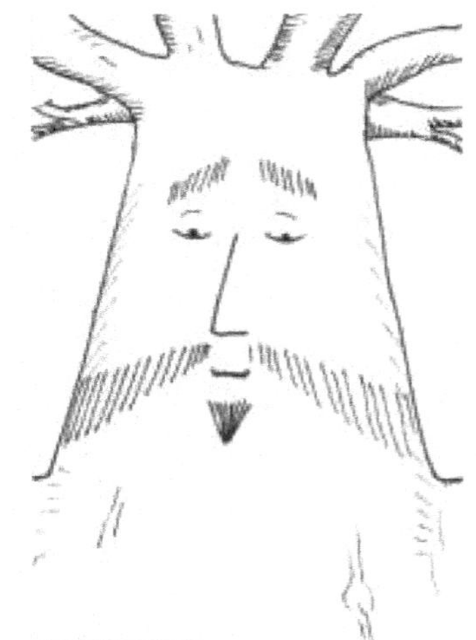

Quercus robur: Thank you for that question. It shows that YOU are thinking! It would be difficult to classify these as "classes" since they are different *types* of thinking. Just by way of review, remember that *conceptual thinking* is a developmental progression initially based on logical thinking (lower mental thought). Through experiences (getting a good grade on a test when you study, or getting punished when you break the law), you begin to learn from your past experiences, recognizing patterns, which then can be used in the NOW to make decisions for the future. For example, if you study today, you will get a better grade on a test scheduled for tomorrow. So, conceptual thinking is the ability to create patterns and recognize relationships in those patterns, which also lead to recognizing higher truths. For example, when you see the same pattern repeat itself over and over, and then all suddenly a different result occurs, then something different has happened, and you need to readdress the pattern to include that something different. How does it relate to the previous pattern? In this way, you create higher truths, expanded knowledge of how things are connected and how they work together.

X-Breed: Of course, we understand that, but isn't that critical thinking?

Quercus robur: Another insightful question. *Critical thinking* provides ways for you to make sense of and evaluate the value of data and information, whether that's based on the information or opinion. Critical thinking IS related to conceptual thinking since in critical thinking we are always searching for larger concepts, larger truths. However, critical thinking is a choice, that is, choosing approaches that help you form your own opinions of information resources based on an open-minded evaluation of the features and components of those resources by using, for example, problem-solving skills. As noted, these skills would include listening, considering other points of view, negotiating, and evaluating one's own mental models, including beliefs, values, and perceptions. Critical thinkers consider not only the data and information, but the context of a resource, questioning the clarity and strength of reasoning behind the resource, identifying assumptions and values, recognizing points of view and attitudes, and evaluating conclusions and actions. And the *Knowledge Capacities* we talked about provide different approaches to critical thinking. Creative thinking is a capacity every single human has. Every time we do something for the first time it is creative! Innovative thinking is when you use that creativity and your knowledge (which is effectively applying information) to develop an effective product or process for yourself or others.

Mouse: But we're not humans!

Quercus robur: Ah! But you REPRESENT attributes that ARE human, helping unleash the human mind.

SECTION III: The Ascent—A Formidable Journey

They were ready. The camp had been broken down and what was needed for the ascent stowed away in their backpacks. There was a sense of excitement, of the unknown. What lay ahead of them? It seemed the more they learned, the less they knew, and several of the critters realized that was an old adage they'd never really understood before. Then the Eagle swooped down with the call to move forward, and the critters moved into a line according to their self-perceived experience, and began the steep climb upward.

As they moved up the mountain, quickly at first, then slowing as the incline increased, thoughts flitted through their minds, sometimes of the more formal learning, and other times of conversations with their fellow Questers. A conversation around patterns that play themselves out in nature seemed particularly "sticky". Struggling to remember their exact names, the Kid searched the passing rocks and vegetation for signs of the triangles and cubes of the Platonic Solids, making up all the mineral substances that are part of the earth's crust. Bouncing along the trail, as the Mouse looked upward contemplating the expanse of the Universe, his small but powerful mind kept considering words shared by Quercus robur, *as above, so below; as below, so above*. What exactly did that mean? And so, with their silent thoughts, they continued upward.

Chapter 6: Self as the Ground of Learning [112-139]

Day 6 – Ascent toward the summit; Camp 2
Notes and Activities

Self is a magnificent quality in humans which moves beyond biological drives and cultural habits. It is the totality of the conscious and unconscious mind, the brain and the body, inclusive of everything that it is to be an individual human. While many people perceive themselves as the roles they play, the identities they take on, these are just the *way we choose* to refer to self. The self has so many facets that there is not one place within the mind/brain/body complex where we could locate self. It is the interactions among *all* of these neuronal patterns, firings and connections—with a focus on conscious thought and choice—that create the self! [112-113]

At the integrated level, the self represents a set of knowledge and beliefs interrelated to physical, mental, emotional, and spiritual capacities. This is a *learned pattern,* not a fixed reality, which we discuss in Chapter 9. The "I" is a unique individual in the world, and each individual considers this world and their place in it from a different perceptual framework, creating and recreating their individuated models as a continuous stream of input comes in through the senses. [118-121]

Historically, humans have thought that who we are is destined largely by our genes. But epigenetics—the study of mechanisms by which the environment influences gene activity—is shattering those myths and revising the belief that we are controlled by our genes. The neural development of our brain is NOT set at birth, and the genome (an organism's genetic material) is much more fluid and responsive to its environment than was previously thought. *Genes are not destiny*. What has been discovered is that gene expression is significantly influenced by the local environment of the cells, which to a great extent is under the control of individuals. This is to say that genes are operating options which are modulated by inputs from the environment such as nutrition, stress and emotions, experience, and learning, all resulting in positive or negative behaviors affecting every aspect of our lives. [113-115]

Let's see … the three areas that are important to consider when exploring self as the ground of learning are the power of our beliefs, the unconscious (which is working 24/7 for each of us), and the fickleness of our memory. And then there is one more.

First, ***what I believe, I can do!*** [115-116] Our beliefs are very powerful and, whether positive or negative, affect every aspect of our lives. *What we believe leads to what we think leads to our knowledge base, which leads to our choices and actions, which determines outcomes.* If we believe we cannot do something, our thoughts, feelings, and actions will be such that, at best, it will be much more difficult to accomplish the objective. If we believe we can accomplish something, we are much more likely to be successful, and *this results from choice, not genes.*

Interestingly, since a person's belief system creates a perspective on the world that is perceived to be correct, this introduces a level of perceived certainty that reduces stress and increases repeatability in an individual's life. Thus, beliefs can reduce stress through reducing uncertainty! However, belief systems can also narrow perspectives and thereby limit learning and activity! [117-118]

Second, ***my unconscious is woven throughout my thoughts and emotions!*** [123] While self is developed through conscious choices, the model of self comes mostly from the unconscious. This model creates the idea of *who* we are and the image we build up of ourselves, and *a great deal of learning occurs in the unconscious mind.* [121] Our unconscious can detect patterns without our awareness, and we may even act on these patterns without being aware we are doing so! *Reflect:* Has this ever happened to you? Have you ever said something and not understood where that came from? Or taken actions that you would consciously not have considered taking?

An interesting aspect of the unconscious is that it never lies but it could be wrong! This may seem odd, but realize that your unconscious is a part of YOU, and lying to self would certainly not make sense. However, what emerges is what is *perceived as truth* based on all the experiences and incoming information from the environment YOU have largely chosen to immerse yourself in and your thoughts related to that environment. In other words, *your unconscious is highly influenced by your conscious choices*.

MORE ABOUT TRUTH and PERCEPTION …

Truth is highly contingent on *a person's level of consciousness and level of perception*. Perception, which is impressions, attitudes and understanding about what is observed, is the result of using our senses to acquire information about a situation or the surrounding environment. Since each person is unique, both in terms of physical, mental, emotional and spiritual makeup, as well as in terms of experiences, individual perceptions also vary. For example, the value of truth—whether subjective, operational, hypothetical or intellectual—is dependent on its meaning as *translated by individual perception*.

As our Sherpa Beneficial Insect reminds us: *What you see depends on the direction from which you look.*

There are many ways to "open the brain" and tap into our unconscious. [124] Since a great deal of past learning resides in long-term memory and the unconscious, it is important to be able to retrieve what we know when it is needed. This "retrieval" may take the form of intuition, heuristics or rule of thumb processes or methods, judgment, or inference. One approach is to put yourself into situations which will trigger potential responses in the direction you desire. Another related useful triggering technique is to learn to ask your self the right questions, and carefully listen to your answers, engaging in conversation with your self. Dialoguing with a "trusted other" is another way to bring out information and knowledge that you may not know you have or to correct information and knowledge that needs changing.

Sound can also play a role in opening the connection between the conscious and unconscious mind. An example is hemispheric synchronization, which is the use of sound coupled with a special binaural beat to bring both hemispheres of the brain into unison. [125] This inter-hemispheric communication is the setting for brain-wave coherence, which facilitates whole-brain cognition and a physiologically reduced state of arousal while maintaining conscious awareness.

X5: What I Believe I Can Do

Our minds are a remarkable mix of past experiences that create patterns of unconscious behaviour and thinking for emotional, mental, and physical confidence and wellbeing. We are also consciously thinking about how to act in different situations and would like to try new things, but are very comfortable with what we know well. We are in a constant state of flux between what we are doing and what we are interested in trying to do. We sense we will benefit from challenging our ambivalence and using the ICALS model to unleash our minds. The unicorn gets it.

To understand what you really can do, you need to open your mind to alternative possibilities.

We envision changing our experiences, opportunities, and accomplishments by creating new options and then creatively stepping outside the restraints of our current reality and finding the path where our beliefs help us to achieve our full potential, stretch our capabilities, and build our confidence. In effect, we move from a mindSET statement of "What is!" to a mindFLEX question of "What is possible?" as we contemplate the future.

This is a shifting from focusing on our assumptions and limited perceptions and how these constrain us, to accepting that affective learning will enable alternative possibilities in the future. The unicorn represents the ability to detach from your perceived reality and challenge yourself to act in a way that your unconscious may have normally blocked. This simple Xercise helps find that path to new possibilities.

STEP 1: Select an aspirational area of your life where you would like to achieve something new and beneficial. As you are successful in this area, it will give you confidence for trying other areas.

STEP 2: What desired outcome would you like to accomplish? How will it add to your life?

STEP 3: What values and beliefs do you have that will make this undertaking possible? Connecting deeply with your values and beliefs will enhance your learning about how to go the distance.

STEP 4: What barriers do you perceive that may prevent you from achieving your desired outcome?

STEP 5: Create a set of actions that will mitigate the barriers you have identified. It is helpful to make these short individual tasks in the format: verb (action word), adjective (descriptive world), noun (subject on which the action is being taken), then include a link to the outcome of that action. For example, from an individual perspective, you might think: Replace (action) negative (adjective to be specific) thoughts (noun) with positive thoughts to create awareness of possibilities. An example in an organizational setting is: Remove (action) old (adjective to be specific) procedures (noun) to ensure any confusion about the process is removed. Then you might publish new procedures and facilitate practice training. NOTE that while identification of barriers (problems/issues) is often necessary in order to discover solutions, the focus is on the solutions. In this organizational example, a simple directive making the old irrelevant and focusing on the new accomplishes this in swift order. It is sometimes more difficult for an individual to "release" the old; however, the best way to accomplish this is to focus on the "new". [133-138]

STEP 6: Dig deep into your knowledge capacities and imagine reframing your outcome to *double or triple the result you originally selected*. Think possibilities. What do you see now?

STEP 7: There is a saying that started on the New York City Subway system to help people think about transfers; some say the folks in Maine started it, others claim it, and some have even sung songs about it. It simply says "You can't get there from here!" Then someone tagged a new ending "But you can get here from there!" In your past experience, how has a clear picture of what you wanted helped you see the path between your outcome and where you're coming from?

STEP 8: Two things: First, what are milestones on the path? Second, what can you plan on happening somewhere in the middle to provide you *incremental success? This will confirm that what you believe is doable* and that you are *Unleashing*.

Third, ***my memories serve me in distinct and powerful ways!*** Your memories are your self's "A Team"! Humans have three distinct forms of memory storage capabilities: *sensory, short-term (including working memory) and long-term.* [126-127] Sensory memory is short, generally a few seconds in endurance. It refers to information received through the senses. Short-term memory takes over when information is transferred from sensory memory to our consciousness. It can engage approximately 5 to 9 bits of information for about 30 seconds. A second form of short-term memory is working memory, which occurs when material is kept in conscious focus for longer periods of time. This happens when you are studying, repeating or rehearsing, or focusing for a period of time on a core issue. Because short-term memory can only handle 5 to 9 bits of information at a time, displacement occurs when it is full and a new bit of information enters. For example, it can be quite difficult to remember a phone number with the last couple of digits dropping out of memory.

Long-term memory is relatively permanent and can be thought of in terms of declarative and nondeclarative. Declarative memory includes factual knowledge such as meanings, concepts, and math, and episodic memory such as events and situations. Nondeclarative memory includes acts and habits which are done by rote or repeated practice and conditioning. An example often used is riding a bike.

Memory—which is involved in all phases of the learning cycle—is not stored in one place, but scattered throughout the entire cortex! Further, the mind does not store memories in the same sense that a computer does. It only stores the most important part of incoming information, that is, its meaning or invariant form which captures the relationships and not the details of the moment. [130] When the memory is recalled, the gaps are filled in (re-created) for the moment at hand, so they are never the same. [129] This means that *individuals cannot fully trust the details of their memory*!

Recalling memories can be difficult. [131-133] Since memories are stored as temporal sequences of associated patterns, information is much *more easily remembered and recalled if put in a story-like form.* Human memories themselves are story-based! The way we remember anything complex is through a series of thoughts or events, with one part of the story related to the next part of the story in linear fashion. Stories come with many indices (decisions, conclusions, places, attitudes, feelings, questions, etc.), that is, multiple ways they connect to previously stored memories. This also occurs in songs, with the patterns of the music and the "story" very resilient to neuron death since 40 percent of the neurons composing a memory pattern can be lost and the pattern still recalled! Our brains can even make sense of what first appears as nonsense! You've probably all seen the Wheel of Fortune game show where letters are interjected one at a time and the winner is the person who guesses the word or phrase first, or had the experience of proofreading garbled words. Based on Cambridge University research, because our brains process all of the letters in a word at once, there are certain patterns that make it easier; for example, having the first and last letters in place. You can have some fun with this! Have a friend take a sentence and randomly scramble the internal letters in a word, OR randomly eliminate 40 percent of the letters, placing the blank spaces throughout the sentence, and see if you can read it. *Game show here I come!*

Oar birans use cnotxet to priecdct what wrdos wuold lgoiclaly come nxet in odrer to from a corheent snentece.

We'll talk more about the amazing predictive powers of the neocortex when we talk about the human skill of planning in Chapter 10. Meanwhile, a few more important "nuggets" about memory. Memory patterns cannot be erased at will; they decay slowly with time when those neurons are not used ("Use it or lose it" applies to memory as well as your body muscles!) [133]

And fourth—and here it comes—**don't forget how valuable your sleep can be!** Less than seven hours of sleep has been found to impair memory. [135] So, forgetting can be both difficult when you want to forget something, and easy when you aren't getting enough sleep!

How does this help you manage your self?

- ***What I believe, I can do!*** When you set your intention in a specific direction, be sure to know you can accomplish this and keep that positive attention focused on your desired outcome.
- ***My unconscious is woven throughout my thoughts and emotions!*** While you don't have full control of your unconscious, it is manifesting based on past input and experiences, your emotions and feelings, and the current condition of your physical, mental, emotional and spiritual bodies. Thus, a first step to managing self and what emerges from your unconscious is clearly ensuring the health of all of your bodies. A second step is recognizing what is emerging and consciously addressing it, that is, through bringing a feeling, thought, or physical reaction into your conscious awareness, you now have conscious choice. The secret here is repetition. What has been embedded in your unconscious most likely occurred through repetition of past feelings, thoughts and experiences, so what is needed is a consistency of new thoughts, feelings, and experiences that become more important than the ones steeped in your unconscious. If you choose to change embedded feelings, thoughts, and reactions emerging from past experiences, then consciously choose new feelings, thoughts, and reactions and consistently repeat those in similar situations. Remember, YOU have the ability to co-create your future; YOU can manage your future through the conscious choices you make today.
- ***My memories serve me in distinct and powerful ways!*** As a learner, there are so many ways to strengthen your memory. Recognizing the power of patterns, create memory keys through mnemonics, setting up memory cues such as creating a word with the first letters of a series you wish to recall. An example is the individual change model used throughout the *Unleashing* book: *AUBFOE*, which reminds us personal change requires Awareness, Understanding, Belief, Feeling good, Ownership and self-Empowerment in terms of knowledge and confidence. Recognizing the power of story, you can weave what you desire to remember into a fantasy, OR into your own life story, of which YOU are the creator!
- ***Don't forget how valuable my sleep is!*** It's so easy to say, "Oh, I don't need that much sleep!" and then move through life in that fashion. Only … you do! Don't try and fool your self in this regard! Pay attention to the physical requirements of life so that your personal mental capacity can fully address the challenges and opportunities of fully living.

Memory Nibs

- Everyday roles are how we choose to manifest self!
- Self is a learned pattern.
- Genes are not destiny.
- My unconscious is working for me 24/7.
- While it may not always be right, my unconscious never lies.
- Memories are fickle. They are re-created each time we pull them up!

Chapter 7: Managing Self [140-164]

Day 7 – Ascent toward the summit; Camp 3

Ah! There's that word "managing" again. [140] It's such an important word when we combine it with "self"! While we talked about this in the introduction, it's worthy of a bit more conversation here. In Unleashing, three aspects of managing self are presented based on the new managerial role in organizations forwarded in the book *Beyond Leadership: Balancing Economics, Ethics and Ecology* by W. Bennis. The first aspect deals with individual preference developed through experience that provides general insights into people, things, ideas, and events of interest. The second aspect is looking at life from the highest possible aspect of the human, thinking strategically both in terms of running the overall system (eating, exercising, sleeping, attitudes, learning, etc.) as well as applying and responding to environmental forces (choice).

The third aspect has to do with an individual's in-depth insights into their own inner dynamics, covering the functioning of the body, mind, emotions, neurosensory system, and states of consciousness. In other words, the self is being asked in this new role of managing its self to engage preferences and insights—as well as professional and managerial know-how—as we move through the various roles of planning, organizing, commanding, coordinating and controlling. This means thinking strategically while listening to the feel of our inner dynamics—our body, mind, emotions and soul (which is defined as the animating principle of human life in terms of thought and action, specifically focused on its moral aspects, the emotional part of human nature, and higher development of the mental faculties). What a lot of thought! We'll talk more about the soul later.

VISUALIZATION AS A POWERFUL TOOL FOR MANAGING SELF

At my height I have a pretty clear view of what is happening on the earth! For the majority of people, vision is the main way they process the world around them. And, because of ever-increasing social media and such, the brains of the younger generations are literally wired for faster programming and more visual programming. Do you know that approximately 70 percent of the neurons in the cortex are devoted to vision, with, interestingly, two out of three of these neurons focused on *inner vision*, that is, imagination, dreaming and such. This ability to visualize—engaging the past, present and future to creatively imagine our desires and dreams—is a great gift to humanity.

This is never the same, of course, because each person is unique in terms of past (heritage, personality, experiences), present (self, needs, passions) and future (desires, dreams), which means that the only person capable of understanding and managing your self is YOU!

Nonetheless, a greater understanding of self in terms of learning, which, whether at the level of cells or self, or anywhere in between, is what enables us to adapt to the environment, can facilitate choices that head us in the direction we desire. [142]

Time for three ideas. The first one is really important having to do with intention and attention, then we are going to consider the idea of presencing, and then explore the value of living and learning in an enriched environment.

First, ***what I intend to do and where I focus my attention create the future!*** [143-147] Intention and attention are tools of the self that are directly related to consciousness. They lay the web for interaction with the world in which we act. *Intention* is a choice of the self, the act or instance of mentally and emotionally setting a specific course of action or result, a determination to act in some specific way. Intention is the structure of action. If we set an intention, it is an intention to *do something*. However, states such as those represented by beliefs, fears, hopes, and desires *insinuate* intention, and they are *about something*. So, as you will notice, actions based on intention can be unconsciously driven, with the intent perhaps noticed after an action was taken!

Attention, a conscious focusing, can only occur in the NOW (the present instant), and sustained attention is a series of NOWs. In one context, the self could be thought of as a hierarchy of goals, because *it is the purpose and goals of the self that focus the largest amount of attention*. Not only *what* we pay attention to, but *how* we pay attention, is important. The frontal lobe helps an individual pay attention and ask good questions; a more developed frontal lobe allows you to take better advantage of new knowledge, to know what to focus on, and to relate it to life experiences so that it has more useful value to you. A more developed frontal lobe occurs through mental exercise. Since there is a direct connection between both mental and physical exercise and blood flow and oxygen to the brain, exercise increases brainpower. In fact, at any age, mental exercise has a global positive effect on the brain. And planning—mental forethought focused on achieving a specific goal—can help clarify and solidify our intention as well as set up a roadmap (actions) that focus our attention. Planning uses patterns from the past to predict the future, and intention is all about the future! We'll think more about planning later on because it is a really important skill.

Mouse: We get plenty of exercise.
Feline: And we're glad to help you out with that!

Amazingly, *stress can focus attention*! [147-153] Of course, there is an optimum level of stress for each individual. Too much and there is NO learning. Just the right amount, that is, that causes a moderate level of arousal, with neural plasticity initiated which facilitates maximum learning. How do you know the right level of stress? That's a hard question to answer, because it is YOURS to answer. Each person is different. Stress is an active monitoring system that constantly compares current events to past experiences. If the emotional content of incoming information is one of strong fear or uncertainty to the individual, stress is created that may significantly limit any learning. Conversely, if there is too little arousal/stress involved, there may be no desire for learning! Excitement is similar. Excitement can serve as a strong motivation for learning, but cannot be so strong that it becomes highly stressful moving to anxiety. Since stress depends on how we perceive a situation, it can be controlled. Just looking from a different viewpoint—perhaps using a Knowledge Capacity talked about in Chapter 5—can change the way we feel about a situation.

Second, ***living in the present allows the future to flow freely into my reality.*** [155-159] Presencing is attending to the moment at hand, that is, being aware of and paying attention to whatever is taking place in the present. Since awareness and attention are two key components of mindfulness, it is clear that "presencing" has a great deal to do with mindfulness! Intent is also necessary for presencing. This is the idea of not only "letting go" but "letting come". Letting come is the process of *allowing*, whose importance cannot be overemphasized. As the adage goes, we can be our own worst enemies! As we realize the power

of our mind/brain in terms of thought and feelings in the process of co-creating, we do not want to interject barriers to our desired progress forward. Presencing opens us to receiving and learning. Despite a focus on self, presencing is not an individual concept, but rather considers the individual as a living system that is part of the whole. In other words, there is a sense of social identity present. In the context of the mind, this would mean *consciously* showing up, being aware of our inner connections to the larger field of which we are a part.

Intent provides purpose to life and living. From a larger perspective, without presence we would become our environment, chameleons who are brilliant at imprinting what we see and bringing it into our own energy. When we are conscious, we are present, and while these are interrelated, the idea of presence goes beyond an awareness of NOW to include deep listening and the ability to through awareness move beyond the way we've done things in the past, that is, it brings with it the freedom of choice.

Third, ***an enriched environment expands my mind.*** [161-163] An enriched environment actually increases the formation and survival of new neurons, affecting both the ability to learn and the quality of the learning experience! Further, an enriched environment can actually produce a personal internal reflective world of imagination and creativity, stimulating the mind to associate patterns and create new possibilities. Only what does that term "enriched environment" actually mean? Certainly, as our Questers can affirm, being outdoors connecting with the beauty of nature is an enriching experience. And when we are in a setting that has lots of interesting and thought-provoking ides, pictures, books, statues, etc., that can be enriching. And music and art, certainly those are enriching? And even people—interesting people—can provide an enriching experience. Yes, yes, yes.

Back before any of our OrgZoo characters were born—who have only been existence for 15 years, although no doubt they were a glimmer in their creator's eye for much longer—we might have said that effective learning required concentration: no physical, mental or emotional distractions or disturbances. But an enriching environment is highly sensitive to the learner, and today's learner thrives on technology augmentation, whether computers, tablets, phones, continuously interacting with others, stimulating their thoughts, capturing their emotions, and accelerating their actions. Is this an enriched environment or is it interfering with reflective thought and conscious choice? Reflect on this as our Questers continue toward the summit and as we continue to unleash the human mind.

How does this help you manage self?

- ***What I intend to do and where I focus my attention create the future!*** What this means is that YOU are in control of your future, and through focusing on something you desire—planning for it, setting your intention, and consistently paying attention to it—you move closer and closer to making it a reality! Engage your creative imagination. What do you desire? What do you want your life to look like in the future? If you can't answer that, then you don't have a direction to move forward.

ATTENTION AND INTENTION WORK TOGETHER

It is worth taking a closer look to see how attention and intention work together. Perhaps we can pull from a graphic included in the *Unleashing* text. What it shows is that responding to your thoughts and feelings, you must continue to have focused attention as you set you plan and set your intention on a specific goal, and then work to achieve it. [144]

Now, note there is reference to a continuous series of NOWs, which is related to your attention. This is because attention can only occur in the NOW, and it requires a continuous series of these in order to stay focused! This can be a real challenge in the chaotic environment we live in today! But if you choose, you can do it! I know you can. Just like we've all made it to 7,000 feet!

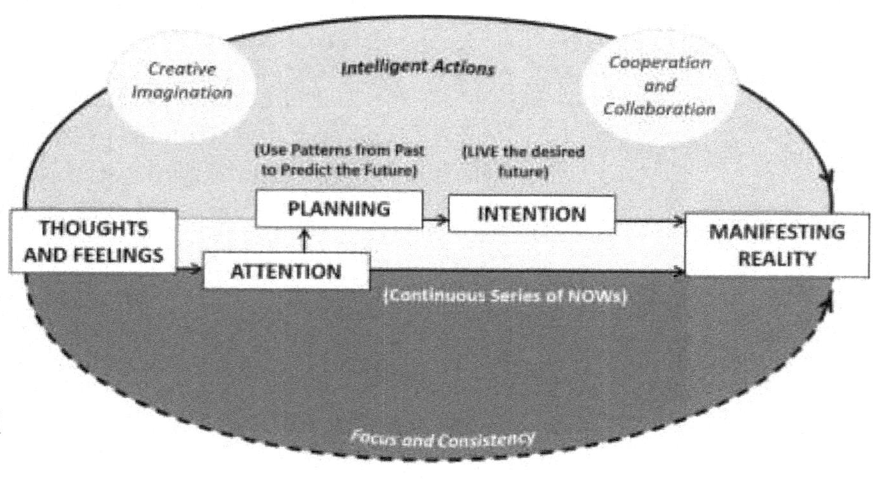

- ***Living in the present allows the future to flow freely into my reality***. When you think about it, there is only the present. The only thing you can change about the past is your perception of it—and as you create the "story of you", you can shift that story a bit to make sense, and that will eventually become your memory, your personal truth. And although the future draws your interest and enhances motivation, the present captures your attention. As soon as the future is here, it is the NOW. What you do in the NOW quickly becomes your past AND affects the NOWs of your future. In other words, what YOU THINK and what YOU DO determines your future AND your past! By managing your self in the now, you are setting your future into motion.

- ***An enriched environment expands my mind***. What are those things that make up an enriched environment for you? Maybe you enjoy sitting in front of your computer, occasionally pausing to look at your window in a beautiful rural or farm setting. Or maybe you enjoy the "Mozart Effect", listening to a particular type of music that either calms (if you're agitated) or invigorates (if you're lethargic) you. Now you know that YOU have a personal optimum level of stress (or excitement) at which you learn, so identify it, and CHOOSE to place yourself into situations where that occurs for optimum learning.

Memory Nibs

- Engaging my insights and preferences is part of managing self!
- A lot of my neurons are focused on inner vision.
- My intention and attention create my future!
- Presencing is paying attention to what is happening now.
- The optimum level of stress can actually help me learn.
- Enriched environments enrich learning!

X6: Compassing

Overview of activity: This activity is useful for reassessing your direction in life. It uses the "direction" metaphor of a compass, considering your directions in the past, present and future. By assessing where you have been and how those experiences have impacted your life (positively or negatively), you can build a plan for the future that pursues your dreams while reinforcing positive aspects and mitigating against negative elements. It can be facilitated with inexperienced participants and mixed groups with mutual trust. It is also useful as a self-development reassessment activity for experienced people looking to reassess their life purpose and future goals.

Activity process: This is your reflective "Life Compass". Use it to assess where you are and where you desire to be in the future. Think about your past experiences and how they impacted you. Consider some of those you have observed in others (such as your role models) that you would like to emulate (or some you may wish to avoid). Try to balance the number between positive and negative experiences to balance your perspectives (ensuring you are psychologically safe to explore these areas). Start by drawing a compass format on a piece of paper.

STEP 1. Mark the future positive experiences in your North East quadrant (upper right) with a dot. The more positive they are the further North. The further in the past, place more toward the West. Follow this same process for the other three quadrants, using a dot to represent past/positive events (upper left of quadrant), past/negative events (lower left), and future/negative events (lower right). Sherpa Beneficial Insect has volunteered to serve as an example of how to fill in these quadrants. See the self-compass example below.

STEP 2. Consider each experience and what you can learn from each experience. How did it come to happen? What did you do that enabled you to achieve this success, or have this difficulty? What could be done to make more of the positive moments and avoid or mitigate against the challenging moments?

STEP 3. Build a pattern of events (at least 3 in each left quadrant), so that you understand what you desire to experience more often and what elements of your life you would rather remove or limit. Mark activities in your two future quadrants that you would like to achieve (North East) or avoid (South East) as you go forward. Reflect on what you need to do to enable this to be achieved. Plan your future steps and activities using these points as compassing guides to set your direction, including who you want to collaborate with and why.

STEP 4. Your "Life Compass" is a reflective tool that encourages you to explore your experiences more deeply, so that you can set your future direction (intention and attention) in a self-informed way. This will assist you to better plan for the future and alleviate some of your stress and anxiety about past challenges, enabling you to positively more forward to where you want to be ... becoming the best possible version of the future you!

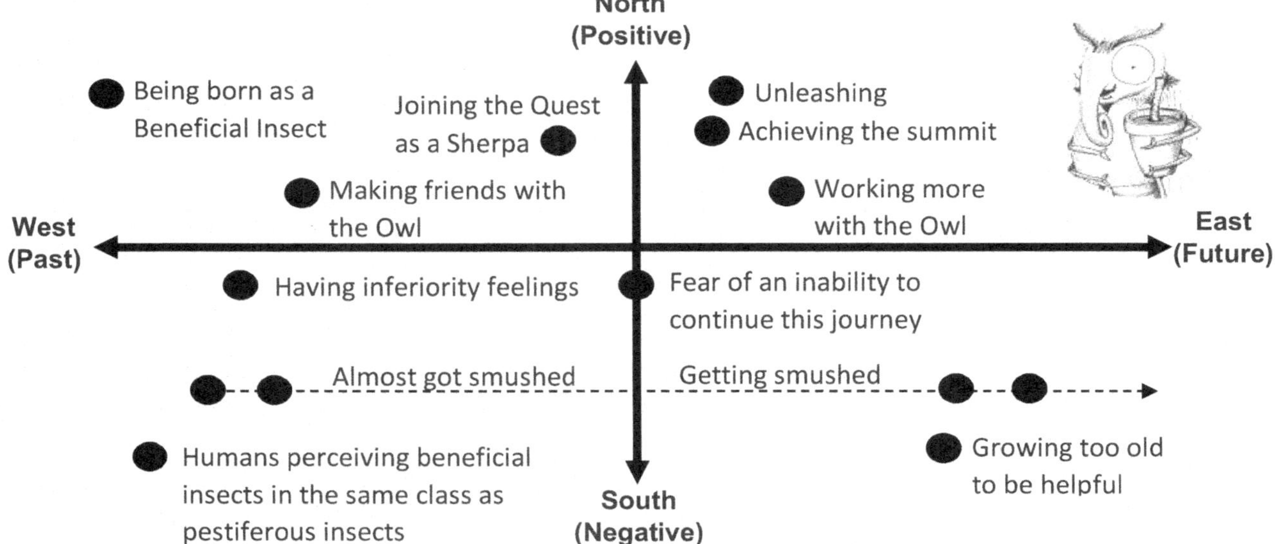

Optional Advanced stage: Compassing can be a very grounding, and yet an "unleashing" experience to continue over time. Although useful as a one off reset activity, it is most powerful as an ongoing self-assessment. This can be made even more powerful if you included a trusted partner into your self-assessment. Hearing what another person thinks about your assessment and planning, can provide you with a perspective that enhances your insights and future experiences. Sometimes we can be too optimistic or too humble in our self-assessments because of sub-conscious biases and fears. Simply- we are not objective about our own self. To get informed independent feedback from caring, trusted others can be powerful for your own insights and planning. If you are in a deep long-term partnership, it is a wonderful experience to do together, to cocreate your joint path forward.

Advice to Facilitators/Mentors (Explain the following to set the context for the participants):

Specific to this activity: (1) Exploring past negative experiences can be very challenging for some people who have previously experienced trauma. Please ensure to create a safe environment. (2) It is helpful to reflect on the behaviours that lead to the successes and challenges. This can assist in the future planning to develop greater behavioural adaptability for the mentee.

General for learning interventions: (1) Ensure you allow plenty of time for self-reflection and exploratory conversation, as this is where the real learning happens. Iterative cycles of reflection over time, using the previous output (your drawn life compass and other notes) as inputs to the next cycle is valuable as it enables deeper and wider reflection over time. As a mentor, it is advisable to encourage your mentee to keep a notebook for this activity that can be kept and redrawn over time as the mentees' priorities change and their experiences and confidence grow. (2) Encourage participants to continue the conversations beyond the activity so that the new learning becomes habit and therefore generates ongoing benefits and stimulates continually learning.

X-Breed: We've never really paid much attention to bugs, and were quite surprised that you were actually a Sherpa on this quest. But we have recognized your contribution, howbeit not at the same level as the Sherpa Prof Owl or Quercus robur.

Beneficial [pausing before answering]: Ah! It sounds like you are much like the humans who perceive all insects as the same! In the distant past—although not so distant for you since the lives of insects are quite short in comparison—I would have taken that personally. But I now know that we all have something to contribute to the greater good, and I was so fortunate to be birthed as a Beneficial Insect who indeed can help others.

X-Breed: We were birthed to create wealth and that's just what we do. However, there seems to be more possibilities in life. And with our education and experience, it seems like we just may be able to do more. Frankly, we were pretty sure we knew everything there was to know about life, only being in the midst of this group of critters on this quest certainly has been eye-opening.

Unicorn [who enjoys staying near X-Breed]: I always perceived you as perfect, just as I am! Or at least, as I thought I was. Are you saying that you are *not* perfect?

X-Breed: Remember that conversation on humility! We took that to heart. **[To Beneficial]** You are right. We each have something to contribute, although some more than others, only it is just logical that if we have more to contribute than others, along with that comes the responsibility to do so. Hmmm.

Triceratops: I'll have to reflect on that.

Chapter 8: Heart-Mind Entrainment [165-187]

Day 8 – Ascent toward the summit; Camp 4
Notes and Activities

Emotions play a vibrant and extremely dynamic role in what it is to be human. Whether you consider them as mental states rising spontaneously or organized patterns constructed out of thoughts and behaviors in the moment, they involve mental and physical reactions to specific situations (people and behaviors). To be sure, they are *long-term drivers that consciously and unconsciously guide our lives.*

Influencing all incoming information, the entire body is involved in emotions and the body *drives* emotions. [165] The amygdala places a tag on incoming information that assigns it a level of emotional importance. If the incoming information is considered dangerous to you, the amygdala immediately starts the body's response, such as pulling a hand away from a hot stove. This is the "fight or flight" reaction, and in early history it was vital to our survival when a wild animal might attack!

In parallel—but much, much slower—the incoming information is processed and cognitively interpreted, although by that time anything of significance to the individual has an emotion or feeling attached to the thought! So, while we've talked about the neocortex as the organ of intelligence, the "new" brain, and as such being objective, it is largely guided by the "old" brain through emotions, which related to passions, values, and virtues, all of which are subjective. In other words, your objective mental thought is embedded with subjective emotions and feelings! The emotions we feel—or choose to feel—affect every part of our lives, personal and professional, whether alone or in a group.

HEART-MIND COHERENCE

As can be seen, the heart and mind have a strong connection. [167-171] Through research from the HeartMath Institute, we now understand the *how* and *why* of that heart-mind connection. The heart communicates with both the brain and the body in four different ways: neurological communication through the nervous system, biochemical communication through hormones, biophysical communication through pulse waves, and energetic communication through electromagnetic fields. The electromagnetic field generated by the heart also radiates outside the body, carrying information that affects the environment and others in the environment. As an information—processing center, the heart affects mental clarity, emotional balance, creativity, intuition, and personal effectiveness. "Coherence" between the heart and the mind is the optimum state for working together of emotions and thoughts, achieving mental clarity and emotional balance.

You've probably heard the expression and felt the rollercoaster of life. This refers to the up and down ride between excitement and disappointment, and emotional highs and lows. Emotions—context sensitive and situation dependent—form a dynamic flow, and humans seem to have a penchant for living from one extreme to another. However, as we ride this rollercoaster, there are choices that can be made, with emotions serving as a guidance system for self, alerting and protecting individuals from harm, and energizing them to action when they have strong feelings or passions, hopefully with those strong feelings and passions consistent with the individual's direction of choice.

Eagle: Together we've climbed to 8,000 feet. No doubt all of us have learned something along the way. Before we turn in for the night, I'd like to have a few volunteers share with the group something they have learned, any new ideas, and since today we focused on the strong connections between the heart and mind, what feelings you are having about what you are learning, about yourself, and about any changes you are seeing in yourself or would like to see in yourself. Take a few minutes to reflect and then we'll start.

X-Breed: How did I end up with you all on an expedition up Mountain Quest II? We could be down in our green valley basking in the sun. It was clear to me from the get-go how challenging this ascent to the summit would be, and I did my best to make it known. Next time, listen to me. This quest is starting to test even my mettle. Now, I understand that managing my ability to learn in new ways changes my self's potential to even exceed my previous accomplishments, which are many. I thank my lucky stars that I'm able to draw upon your determination, and perhaps listen to Sherpa Quercus robur to be on the alert for unexpected shortcomings that even X-Breeds can experience, although these are rare!

Jackal: Now, that's something I never thought I'd hear you say! In fact, I've been pleasantly surprised over the last several days as I've watched an inner strength emerge from several of you who I didn't expect to successfully achieve this quest! So, regarding feelings, I'd have to say that mentally watching that strength emerge gives me rather warm feelings towards those individuals. Quite unexpected!

Feline [to the side, not aimed at anyone in particular]: Even the mouse is showing some promise!

Kid: When you think about it—and I'm starting to do a lot of that these days—this whole quest is part of the rollercoaster of life!

Beneficial Insect: That rollercoaster of life idea means more than going up and down the path toward the summit. When you're riding a rollercoaster you have to just surrender to the ride. All the turns and highs and lows are exciting. And there's part of life where we do have to surrender, and while there's joy in the ride, we also know that our emotions and feelings are ours. We'll talk more about that in a little while.

Dog: I've noticed a throbbing my head that matches the beating of my heart! So that must mean something really good. And even when we're struggling up a steep cliff, I'm really happy … happy we're doing it together, happy I'm not falling, and happy when we make it! So, I'd say I'm just really feeling happy!

Triceratops: You're always feeling happy! And seeing that I've started reflecting about the heavy thoughts and feelings that are always with me, and have been watching how the gibbon, who also always seems happy, swings around. I can't swing. So, I'm trying to figure out how to lighten up the load I carry. I'm imagining a big "sucker" bulb that I can use to pull it out, but maybe that won't work. Or swimming in a river to wash it away, but that only handles the outside, not the inside. And, then, maybe some thoughts and feelings that make me powerful would disappear as well. I'll just have to keep reflecting.

Everyone on this quest has had their own personal experiences with emotions and feelings! Let's see if we can bring some of that into context, perhaps expanding our understanding not so much about the power of emotions but focused more on their role in our lives. Our three focal points are the relationship of emotions and learning, the relationship of truth and emotions, and managing our emotions.

First, ***my emotions can accelerate or hinder my personal learning***. [171-173] Emotions can increase or decrease neuronal activity, not only influencing awareness, but also the ability to focus attention. Developing our passion for learning opens our mind, creates possibilities, and expands the scope of interest and creativity. Avoiding learning closes the mind to options and causes it to focus on safety and protection, which minimizes learning. When a learner manages their emotions by choosing to take a positive attitude, they will *enhance their learning*. Thus, while the external environment and the material to be learned may have a significant role in the learning outcome, *the perception and interpretation by the learner significantly impacts the learning results*.

In the NOW, our unconscious mind is continually *interpreting* situations and these interpretations influence our emotional response. A fascinating aspect of the emotional system is that it is not concerned with the *details of information*. Its primary concern is with the *meaning and impact* of the information! [173-176] This is highly individualized and subject to circumstances. Events themselves, that is, things happening in the environment in which we live and interact, *do not themselves have meaning and impact for the individual other than what they assign*. It is our self, emerging from all the experiences of life, that assigns meaning and impact to events. In other words, WE decide what things mean to us! What a huge responsibility that is.

Not surprisingly, the *amount* of meaning in life is directly related to an individual's level of consciousness. And levels of consciousness are deeply related to emotions and feelings! This idea of levels of consciousness keeps emerging, so perhaps we should ask our Sherpas: exactly what are these levels?

LEVELS OF CONSCIOUSNESS

In D. R. Hawkins research, levels of consciousness are correlated with emotions, perceptions attitudes, worldviews, and spiritual beliefs. Ranging from 0 to 1,000, the progression is:

20	30	50	75	100	125	150	175	200	250	310	350	400	500	540	600	700-1,000
Guilt	Apathy	Grief	Fear	Desire	Anger	Pride	Courage	Neutrality	Willingness	Acceptance	Reason	Love	Joy	Peace		Enlightenment

The levels falling under 200—which are weak attractors and represent negative energy—are focused inward, centered around self. For example, consider level 150 (anger). We always say, "I'm angry at …" or "I'm angry because of …" and give credit to external sources. However, the anger is with YOU; our emotions are *our* emotions. As you move up the levels of consciousness scale, your increased awareness offers greater choice as life unfolds. We are no longer victims to our environment in terms of emotional responses, but can use those very human emotions as the action guidance system they were intended to be. [175]

Second, ***my truth is grounded in my emotions.*** [176-177] Truth—a living, dynamic awareness that grows in its meaning and value as we learn and our consciousness expands—is a shifting target relative to our self. [176] The value of truth is dependent on its meaning as translated by individual perception, all very much entangled with emotions. And because we are complex adaptive systems living in a continuously changing reality, an apparent truth (level of observation) and imagined truth (level of distortion) may well accompany a perceived actual truth, which is, of course, only actual for the moment at hand and in terms of our perception!

In our learning experience, humans are truth seekers, with our individual truth locally structured based on our perception, prior learning, and life experiences, with the value of truth is *dependent on its meaning as translated by individual perception*. Because truth is structured, it can be discerned in comparison to another truth and can be given a relative value regarding the level of truth, that is, a mathematical representation of how much a conceptual truth is true in a specific example. So, this relationship of emotions and truth can be shown mathematically, which affirms *the transfer of emotions and intentions along with thought*. [177-178]

Another related concept is that of resonance. Resonance is a quality—a richness or significance—which evokes an association or strong emotion. All objects have simple wave functions that have unique wave signatures with a natural vibrational frequency. A resonance of thought occurs when written or verbally expressed ideas are consistent with the rhythm of your natural frequency, that is, vibrating at the same frequency. Inducing resonance is a result of something external resonating with internal information, *bringing it into conscious awareness*. This resonance amplifies your feelings, which often are then expressed as emotions. [179-180]

Third, ***I can learn to manage my emotions***. [180-187] Considering the large role emotions play in our lives, it sure sounds like a good idea to learn how to manage them! And, you can do just that. The very first knowing is to *recognize that they ARE ours*, remembering that although we often perceive emotions as caused by external events or people, they emerge within ourselves.

A second knowing is *an awareness of those emotions and what they are communicating*. They are meant to be a guidance system to ensure that we are in

Yes, our emotions are OURS! Let me tell you a story that shows this is true. Imagine yourself as a young woman who has the opportunity of a lifetime to interview with a large public relations firm in New York City. You meet with several executives, you provide writing samples, they look at some of your layouts. You get the job at twice the salary you expected! Floating with happiness, you exit the office building and head toward your car. It is dark, and you notice two large men walking behind you. You speed up; they speed up. You turn the corner; they turn the corner. You feel a level of fear. You start running; they start running. Catching up with you, one of them reaches out and grabs your elbow, saying, "Excuse me, miss. You dropped your wallet."

Clearly, the feelings of fear were based on expectations driven by internal beliefs and perceptions, not by reality. The emotions building within you were YOUR emotions, your imagining of possibilities. Nothing bad or negative happened other than in your perception of the situation, which is driven by the "old" brain whose job for millions of years has focused on survival. While you cannot run away from your fears, acknowledging them in context as a signal that you are out of alignment puts them in another perspective. Our emotions are our guidance system.

alignment with ourselves. Thus, when we are feeling joy and love, we are in alignment; when we are feeling anger and fear, we are NOT in alignment!

When negative feelings occur, it is time to do something differently. As a rule of thumb, it takes only 17 seconds of focused feeling to shift from one emotional state for another. You do this through shifting to good-feeling thoughts (memories, dreams, desires). In the body's complex responses to emotional arousal, there is an electrochemical pathway that moves from the brain through the limbic system and then throughout the body utilizing adrenal glands and the autonomic nervous system. By shifting your thoughts, refocusing your attention, you can change your emotions.

A third knowing is to recognize that *internal emotions are triggered by external situations and events*—which can consciously be avoided or, conversely, set into motion. Physical bodily changes follow our perceptions, and the feelings that emerge ARE based on emotions. Feelings as a form of knowledge have different characteristics than language or ideas, but they can lead to effective action because they can influence actions by their existence and connections with consciousness. When feelings come into your conscious awareness, they can play an informing role in decision-making, providing insights in a non-linguistic manner and thereby influencing decisions and actions. Listen to your feelings, use your mental faculties to understand what they are telling you, and then consciously choose your course of action.

A fourth knowing urges us to *release emotions no longer in service*. Within the scientifically measurable energetic fields of which we are a part, energy—both in the form of thoughts and emotions—can be captured, slowed or stopped, both intentionally and unintentionally, consciously and unconsciously. This means that you may be bogged down by memories or feelings from the past. Blocked or stuck energy is in stasis, that is, a motionless state. Remember, a complex adaptive system cannot exist long in stasis, but must be either growing or declining, expanding or dying. In the body, stuck energy might mean preventing fluids from flowing normally through their regular channels (a blockage to air circulation caused by asthma or food lodged in our airway) and lead to death. Thus, learning how to "let go" of negative emotions and thoughts is critical to continued growth.

A fifth knowing is that the brain can generate molecules of emotion to *reinforce what is learned*. Through positive thought and attention, this phenomenon can provide control over attention during the concrete experience.

What does this mean to managing self?

Emotions are our guidance system, and learning how to manage them is essential to our health and continued growth. Holding onto negative emotions can hinder learning and affect our health. Creating positive emotions through engaging in positive learning experiences not only accelerates learning and physical well-being, but helps us bring greater meaning into our lives.

Humans have the ability to balance their emotions. Thus, once emotions are acknowledged, embraced, and managed, meeting their purpose and responsibility as a guidance system, it is our choice how and to what extent we continue to "feel" and be governed by them.

Memory Nibs

- Our emotions affect every part of our lives!
- The heart and the mind have a STRONG connection.
- Events don't have meaning to me other than what I assign!
- My emotions are MINE, triggered but not CAUSED by external events or people.
- For my health, Need to let go of negative stuff!

X7: Releasing Emotions Technique

The guidance system which punctuates the positive and negative aspects of our lives, our feelings and emotions needs to be honored, and, as appropriate, considered in the making of our day-to-day decisions. As positive emotions such as love and joy flow through our lives, they become more meaningful. Conversely, while we may choose to hold on to strong negative emotions to purposefully create a force to propel us on a course of action, once they have been honored in terms of recognition and understanding, it is indeed a good idea to release emotions when they are no longer needed. This simple tool works for that purpose:

STEP 1: Recognize and name the emotions you are feeling, fully acknowledging their presence.

STEP 2: Put your arms around yourself and, rocking in a motion from left to right and back in a self-embrace, and with a feeling of gratitude for these emotions, *say out loud*: "I am having a human moment."

STEP 3: Stay with these feelings as long as necessary, and when you are ready—ensuring that you have learned all you need to learn from their presence—thank your emotions for their presence and for this learning.

STEP 4: Using your creative imagination, choose to release these emotions, visualizing them floating away in a balloon, or imploding into the air, or sending them to a junk yard for potential reuse. Have fun with this. The only limit is your imagination.

STEP 5: When a negative emotion departs, it leaves a clear space that needs filling. To fill this space, spend a few minutes thinking about some happy memories, or engage in an activity during which you feel joy.

NOTE: It's useful to create a small card (wallet-size) with positive thoughts. On one side, write five things you have done and enjoy doing. On the other side, write down five things you have done for others that have brought them joy. This card can be used to spur happy thoughts in the technique above, or can be used to raise your vibration anytime you feel an emotional low.

As the critters listened to this Xercise, several of them wrapped their arms around themselves and were rocking from left to right and back with smiles on their faces. The Triceratops was the first to break the silence.

Triceratops: I'm going to volunteer to "tri" this first. [He looks around to see if anyone is laughing.] Do you get it? "Tri", that's T-R-I, like my name. I'm willing to "tri" this! [He's somewhat exasperated by their non-reaction to what he thought was a pretty good joke. So, he closes his eyes, wraps his arms around himself, and rocks back and forth. Then speaks again, pointing upward.] There all those negative emotions go! See them rising up into the air? There! [He points to a wisp of fog that seems to have settled above them. Now, there is some laughter. Triceratops looks somewhat proud, and speaks softly but somewhat sullenly to the Gibbon who is sitting beside him.] Well, they certainly didn't get the "Tri" joke you suggested I use, but I did get them to laugh, and that is a first.

Gibbon: It might help if YOU laughed, or at least smiled a bit.

Triceratops: It is harder than I thought to make a joke, and I didn't realize that I have to laugh at my own joke! I'll have to do more reflection on this.

Chapter 9: The Extended Reach of Self [188-208]

Day 9 – Ascent toward the summit; Camp 5
Notes and Activities

Humans are social creatures, and the environment is a critical part of and actively engaged in the learning process. Our brains are linked together, and physical mechanisms have developed in our brain to enable us to learn through social interaction. This includes emotionally attuning to others, recognizing their intentions, understanding (or at least trying to understand) what others are thinking, and purposefully choosing how we want to interact with others. The mechanisms for this capability come from adaptive oscillators, which are part of our physiology, and also through mirror neurons which Sherpa Prof Owl will tell us more about. These oscillators are created by stable feedback loops of neurons, and can bring an individual's rate of neural firing into sync with another individual, which is when two people relate well to each other and learn to anticipate each other's actions. [188-189]

What we can learn from others should not be underestimated. Social learning includes not only engaging in our own experiences with others but *replicating the experience of others as our own*, embedding it in our mind and body, and acting accordingly! For example, it's not surprising that an important aspect of learning can be watching someone else doing what you want to learn. What we see, we become ready to do. Let's ask Sherpa Prof Owl to tell us more about that replication and specifically mirror neurons. [194-195]

THE PHENOMENON OF MIRROR NEURONS

Our "self" is in continuous interaction with the environment, largely in the unconscious, and that interaction works both ways! Not only does the observer affect that which is observed, but that which is observed affects the observer. One element of this is mirror neurons, a recently discovered phenomenon in the brain. It was discovered through research with Macaque monkeys, but it is the same in the human brain. The same neurons fire in the brain of an observer as fire in the person performing an action, *providing an internal experience that replicates that of another's experience*, thereby experiencing another person's act, intentions and/or emotions. There is a caveat; this only occurs when the goals of the performer are understood by the observer, creating a resonance between the two. We talked about the importance of resonance at our previous camp.

This process of mirror neurons [194-195] facilitates the rapid transfer of knowledge and facilitates a "neural resonance" between the observed actions and the executing actions. This means there is a positive, reinforcing relationships between two people interacting with each other. This also sets the stage for an existential experience, which will be part of our focus at our next campsite.

Related to the mirror neuron phenomenon, we now recognize that through mental reliving and reflection, we recreate the feelings, perspectives and other phenomena that we observe. In other words, we may understand other people's behavior by mentally simulating it. Empathy is the concept of experiencing the inner life of another, which has been described as the entanglement of resonance, attunement and sympathy. The physiological basis for empathy is normally inherent in brain function, and it plays an important role in both the experience of self AND in the ability for us to distinguish our self from others. The research suggests that through feelings there is an active link between our own bodies and the minds and the bodies and minds of those around us. Thus, the feelings that we each perceive in the course of our daily living may be affected by, *or even belong to*, those around us.

Okay. Let's address three focal points regarding social engagement. How about addressing social attunement, co-creating our reality, and a little about managing our relationship network and the importance of doing so.

First, ***I seek attuned others to learn.*** [190-194] Building on our earlier discussion of emotions, a positive social relationship, and a corresponding environment, support individual learning through the creation of internal perspectives and feelings that facilitate pattern generation and the development of meaning. For example, the biological impact resulting from interactions with an attuned other can significantly enhance learning. During this process the neurotransmitters in the frontal cortex of the learner are stimulated, which leads to increased brain plasticity and neuronal networking.

What is an attuned other? Attunement infers being in harmony with or resonating with another. There is that resonating word again! It's a pretty important concept. Attunement is a shared state, a vibrational entrainment or resonance—inclusive of understanding another's needs and feelings—that opens the door to learning. If it is to learn, the brain needs to seek out an attuned other! Reflect back in your life and identify individuals with whom you have been affectively attuned. The limbic system, the primitive part of the human brain, and in particular its amygdala, is the origin of survival and fear responses. Affective attunement alleviates fear, which we have identified as an impediment to learning. Not only does affective attunement offset the discomfort of the developing complex neural patterns, but it actually attributes to the evolution and sculpting of the brain! So, when you have a relationship with a mentor, coach or another significant individual who is trusted and with whom you are capable of resonance, the dialogue between you is greatly helping you understand, develop meaning, and anticipate the future with respect to actions.

Mouse: This idea of resonance makes sense. WE have a resonance of sorts! We're both very individualistic, very agile, and we energetically exercise together playing the cat and mouse game. Is that affective attunement?
Feline: I don't think so.

Second, ***I am co-creating reality with others***. [199-201] We as a humanity are in a continuous cycle of co-creation such that every moment offers the opportunity for the emergence of new and exciting ideas. Note that in every aspect of creating and learning there is a "co" element. This is because ideas do not happen in isolation. Some of the greatest meaning of life come from sharing knowledge that facilitates the creation of new ideas and potential innovations. Both consciously and unconsciously, as we interact with others, we develop a deeper understanding of others and ourselves and an appreciation of diversity, creating a collaborative advantage. We are able to gain the advantage of other's thinking at or above our personal level of thinking, while simultaneously creating in a way that is uniquely ours, concurrently individualized and one.

Humans are experientially engaging a changing landscape in a continuous cycle of learning and expanding. It is the context of the activity or situation at hand such as a need or challenge that triggers putting things together in an unusual way to create something that may be new and potentially useful (innovation). Humanity is in a continuous cycle of knowledge creation such that every day offers the opportunity for the emergence of new and exciting ideas, all waiting to be put in service to an inter-connected world. And conversations—with ourselves and with others—is often where that learning occurs. Conversations really matter.

Third, ***through managing my relationship network—who I converse with—I am managing my learning*** and future opportunities. [202-204] At every point in the journey of self we are in the process of becoming something more or something else, with vastly different perspectives emerging from one activity to the next as to where we are at any point in time! And what is known to one is unknown to others, or perceived very differently. What seems real to one person can be totally unreal to another. Your truth is considered just a perception by others, and vice versa. [201]

Considering the continuous social interactions, both on the conscious and unconscious levels, it makes sense that our everyday conversations lay the groundwork for the decisions we will make—and events that will occur—in the future. Since time is a scarce resource, it is critical to choose our relationships and interactions wisely, and it is each of us who chooses with whom we interact. There are given basic concepts that help us successfully manage our relationship network. These are interdependency, trust, openness, flow and equitability, all of which overlap. For example, interdependency includes a state of mutual reliance, confidence, and trust. Trust is based on integrity and consistency over time. These qualities facilitate the free flow of data, information and knowledge among people, across teams and organizations, and around the world. Equitability in terms of fairness and reasonableness means that all those involved in the sharing gain something of value out of the relationship.

It is our interactions with others that help prepare us for an unknown future. Through those interactions we have the opportunity to observe and understand different attributes that others exhibit that we may wish to learn and apply as our own. And since each individual is "one of a kind", there may well be knowledge or capabilities within your network that can assist you in the future, and you may well have knowledge and capabilities that can assist others. As can be seen, through our interactions and conversations, we extend the reach of self.

What does this mean to managing self?

As we know, learning is part of the social experience of life, with every day affecting in some way—shifting, changing, expanding, reducing—the complex adaptive learning system that is the human. The overall environment, a trusted other, and the conscious and unconscious state of the learner all have a role in the final effectiveness of the learning that occurs. By being aware of these factors, you may be able to change your environment, improve communications with others, or perhaps adjust your own internal feelings and perspectives to maximize learning.

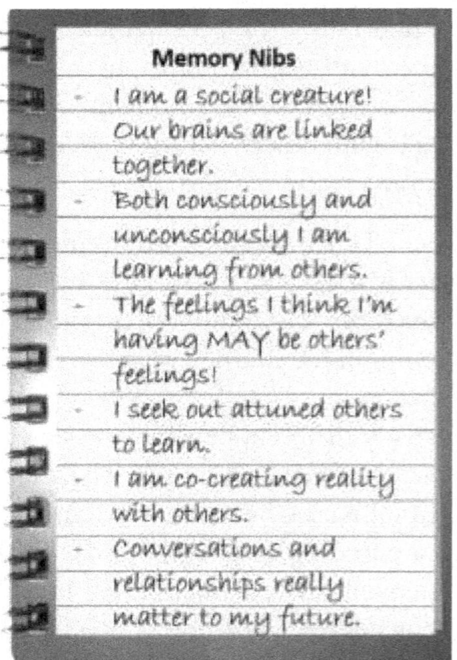

What is quite clear from this neuroscience-supported understanding—looking from the inside-out—is that we are all in this together. Just like we're on the Quest to unleash the human mind together. Although each self is individuated, no self is separate from its environment. And our environment has much to do with perception and choice. We are each embedded in a social network of learning, co-creating and adapting to the emergent challenges and opportunities in the flow of life.

X8: Conversations that Matter [281-284]

Images can stimulate conversations that matter. Sharing our experiences and stories and listening to the experiences and stories of others not only clarifies self-understanding and increases empathic appreciation of others but, because our mind is an associative patterner, triggers new thoughts and connections. The simple process is to show the image or object in question, and then ASK a question. It can be open-ended such as "Tell me what you think about this image?" Or it can be somewhat leading to get a different focus such as "Where do you think our organization fits into this image and why?" When you can, it is a good idea for each person to write down a few quick bullet points, but you can do this conversing as well. You want to capture initial FEELINGS about the question (coming from the old brain) before the thinking mind (the neocortex) begins to over analyze.

Engaging intelligent rules of etiquette for dialogue, start that wider conversation. There is an amazing set of themes that come out of such conversations. Some people inherently see the pessimistic side of their situation and highlight barriers to progress. And some do the opposite and talk of the positives, perhaps even over-estimating the quality of what is being done. The key is to engage participants in exploring the reasons behind the differences to share why there are multiple perspectives. This is where the insights come from as ideas shared stimulate others to respond and new knowledge is co-created.

Combining a creative and out-of-context stimulant with a provocative question and open and inclusive facilitation generates optimal outcomes, leveraging the diversity of views of engaged participants to create new knowledge and insights, which form new options, new ideas. Synergies emerge from connections between thoughts and ideas and emotions, with each component critical to the richness and success of the interaction. The key to remember about such interactions is to not lead the participants to a predetermined outcome, but to co-create a set of thoughts that did not previously exist, and then intermix these to generate a range of options to potentially co-create a future that does not yet exist. The steps for successful conversations are (1) acknowledging differences; (2) respecting other's perspectives; (3) conversing to explore; (4) inquiring to learn; (5) adapting to build capabilities; (6) becoming more capable; (7) engaging in collaboration; and (8) co-creating social value.

Experience has demonstrated the power of exploring individually, and then in pairs or small groups, before engaging in whole group wider discussion, which maximizes the range of perspectives available through divergent thinking before convergent dialogue begins. Cycling between divergent (focusing on openly exploring and option generating) conversation and convergent interactions (focusing on merging, prioritizing and reducing ideas and concepts) enables us to optimize the outcomes.

Gibbon: Hey, Uni, the kid, dog and I are going to try out this conversation that matters idea. Want to join us?

Unicorn: Every conversation I have matters. And right now, I'm tired, so conversing with you is not something that matters. I'm turning in for the night.

Dog: We can do this without him. Only, we don't have a picture to use.

Kid: I keep having a picture in my mind about the sloth falling. But that's a scary picture, not inspirational.

Dog: I keep having a picture about the lion showing up in time to save the sloth's life! That makes me happy. That's inspirational.

Gibbon: It looks like we have the picture in our minds to guide our conversation. You know, I've never liked the lion much, but his strength and courage sure did come in handy in this situation!

Dog: As much as I would have liked to help, I couldn't have caught the sloth.

Kid: I would have been afraid.

Gibbon: But afterwards you weren't. I saw you keeping an eye on the sloth during campfire.

Kid: I just kept noticing. Did you notice?

Dog: Notice what?

Kid: That the sloth stayed awake throughout the whole campfire, and that he was holding his head up straight and looking around. He never does that. What do you think he was thinking?

Gibbon: I did notice that. You know, I've watched some of my tree swinging friends fall a couple of times, and it seems to change them.

Dog: Maybe he was thinking how lucky we are that we have a lion in our group, even if we didn't know he was following us up the mountain. It might be that the lion understands how important it is to be loyal.

Gibbon: So, what have we learned from the picture in our mind?

Kid: I know I'm going to be more careful climbing the rest of the way to the summit. And I'm thankful that the lion is joining us on the rest of our quest.

Dog: And I've seen how the value of loyalty can be a powerful attribute, and that we can trust the lion.

Gibbon: Now, this was only one experience, so we do need to be careful not to carry that learning over to everything we do! There are other things beyond loyalty that count, but you are right, in this situation, it was a very valuable attribute. And certainly, in this situation, the lion was trustworthy. I think that what we are learning is how to think critically.

Dog: Yes, critical thinking.

Kid: Me? Wow! Me critical thinking? This really is a conversation that matters!

10: The Expanded Human [209-240]

Day 10 – Ascent toward the Summit; Camp 6
Notes and Activities

The soul represents the animating principle of human life in terms of thought and action, specifically focused on its moral aspects, the emotional part of human nature, and higher development of the mental faculties. [209-211] From a philosophical viewpoint, it is the vital, sensitive, or rational principle in human beings. So how does that relate to the mind/brain? We've described the brain as the atomic and molecular structure and the fluids that flow through it, and the mind as the patterns created in the brain's 86 billion neurons and their firings across neurotransmitters and trillions of synaptic interconnections. *But can all of our thoughts and feelings be explained by these descriptions?*

It is important to realize that the spaces between neurons (the synapses) direct signals using ions (parts of atoms), which function according to the rules of quantum physics, not classical physics. This means that we are connected to a Universe of probabilities, not laws alone, a field that forms the ultimate foundation of self. In this field there is what is called a quantum Zeno effect. It was discovered that when an unstable elementary particle is continuously observed it does not decay, while when unobserved decay occurs. Since the mind/brain is a quantum system, when you focus on a specific idea, you are holding the pattern of connecting neurons in place, that is, *the idea does not decay* which would occur if it were ignored. This is driven by YOUR thought, YOUR free will. A physical connection to the brain is not present, and the neuronal firings that result from your thoughts and feelings certainly form the pattern, but do not sustain it. YOUR intent, YOUR attention, YOUR attending sustain the pattern. From this point of view, we move closer to understanding "soul".

The "moral aspects" of the soul are the result of learning experiences. They are concerned with what is perceived as right—beliefs or standards of behavior about how to act, largely held in the unconscious as part of the model of self, which create the idea of *who* we are and the image we build up of ourselves. All this is being updated continuously through the cortical columns of the neocortex as sensory and information and knowledge input occur during the course of living. Moral aspects can also be part of cultural norms, or part of an individual's intentions and emotions, and as such can be actively transferred through the cognitive mimicry process of mirror neurons when the goals of the others carrying out actions are understood, creating a resonance between the sender and the receiver.

Virtues are *spiritual characteristics in form*. [214-216] Virtue—morally good behavior or character, the good that results from something—are synonymous with morality, and they are concepts that when thought about expand consciousness. In Phaedrus (24) Plato talks about the soul soaring up to heaven in order to behold beauty, wisdom and goodness. In his "ladder of love", Plato also talks about the beauties of the earth which are to be used as "steppingstones" as we move from fair forms (the human body) to fair practices (organizations working together) to fair notes (learning) to fair living. We are here in this world experiencing (acting and interacting), and through this process learning and expanding, with thought beauty and beauty in thought, and finally recognizing and accelerating to the beauty of life.

X9: The Relationship between Learning and Spirituality

> REMINDER: *Spirituality* is defined as of or pertaining to the "soul"; and *soul* represents **the animating principle of human life in terms of thought and action**, specifically focused on its moral aspects, the emotional part of human nature, and **higher development of the mental faculties**.

Learning occurs by and within complex adaptive systems. Spiritual learning by definition is the process of elevating the mind as related to intellect and matters of the soul to increase the capacity for effective thought and action. As a way of learning, it is involved with the sevenfold spectrum of knowledge: information, sense-making, understanding, meaning, anticipating the future, intelligence, and wisdom. Further spiritual learning is neither dependent upon nor independent of the learner, but interconnected with the learner; it is primarily concerned with long-term learning; and it deals with the non-material and through the non-material affects the material. Note that the human mind has access through the five senses and beyond the five senses—whether originating consciously or unconsciously—to both that which is material (in terms of the physical world) and that which is non-material. This Xercise helps us better understand the relationship between learning and our multidimensional self.

There are three columns next to the 26 words/concepts below. Beside each word put checkmarks in the columns as you feel appropriate. ***Column A*** includes concepts that you feel will improve your life experience. ***Column B*** includes concepts you think can aid your learning experience. ***Column C*** represents those concepts you believe are critical to your inner being, that is, spiritual in nature.

	A	B	C		A	B	C
Abundance				Joy			
Aliveness				Love			
Authenticity				Morality			
Awareness				Openness			
Caring				Presence			
Compassion				Respect			
Connectedness				Sensitivity			
Consistency				Service			
Eagerness				Tolerance			
Empathy				Unfoldment			
Expectancy				Values			
Grace				Willingness			
Harmony				Wonder			

When you have finished this Xercise, scan the similarities and differences, and reflect on those similarities and differences. For those concepts that have checkmarks in all three columns, ask yourself: *Are these the concepts that are the most important in my life?* For those concepts that have no mark in column A, ask yourself: *Could embracing this concept improve the quality of my life?* For those concepts that have no mark in column B, ask yourself: *Could embracing this concept improve my learning?* For each of those concepts that have no mark in column C, ask yourself: *Could embracing this concept offer a greater opportunity to be of service to myself and others?*

If you discovered relationships between learning, spirituality and the quality of your life experience, you are not alone. In a 2008 MQI research study, a direct correlation was discovered between learning and spirituality, which in turn correlates to an improved life. Spirituality suggests a commitment to the process of inner development, a life commitment to growth, and spiritual learning supports personal growth, helps us understand reality, helps others, and contributes to the greater good. Clearly, the material and non-material are married in the conscious and unconscious learning of the human mind.

There is so much we could focus on regarding the inner realm of the soul, and perhaps the need for inner reflection is our first focal point. Then we'll describe the Brain-Heart/Mind-Soul Learning Continuum, and talk a bit about the special human abilities of synthesizing and planning.

First, ***my mind requires time for inner reflection***. [216-220] Recognizing that thinking is mostly unconscious with the unconscious brain always processing, research shows that the brain uses 80 percent as much energy when sleeping as it does when awake. Under deep anesthesia electrical activity is abolished and the metabolic rate of the whole brain is reduced by 50 percent. These findings suggest that the unconscious mind uses about 30 percent of the brain's energy for maintenance during sleep, and that when the mind is quieted while in the waking state—such as through meditation and mindfulness—there is up to 50 percent of processing capacity available.

Recent research supports the theory that there are two alternating attention systems, one focused outward on goal-directed tasks, and a second in a state of rest, a time of inner reflection, considered the default mode of operation. These are codependent systems that coregulate one another, with a direct correlation between increasing engagement in one and decreasing engagement in the other. And research has been able to effectively relate the quality of brain activity during rest to the quality of neural and behavior responses to information coming in from the environment when awake! In the "at rest" state of meditation, the mind is not idle, but rather there is the free flow of thought full of memories, imagination, and future planning, non-attentive but awake mental states, as well as self-awareness and reflection, feeling emotions about events and others, and making moral judgments. The ramifications of this are meaningful for learning. Time to quietly reflect and daydream is *critical for social-emotional well-being*—including development of moral emotions and skills—as well as for the ability to apply attention to task at hand and connect decisions and actions to future goals.

Unique combinations of sound frequencies facilitate various states of consciousness including quieting the mind. Music and the human mind have a unique relationship, with much research questioning whether the human brain may very well be hardwired for music. [236-240] This is not surprising since sound is vibrational with different combinations of frequencies inciting reactions across the physical, mental, emotional, and spiritual dimensions. Note that in fetus development, the ear is the first organ to develop to its full size (which happens around 18 weeks), during life hearing through the auditory nerve has 90 times greater range than our eyesight, and hearing is the last sense at death. For these reasons, the ear has been called the organ of transcendence, directly linking the ear to compassion and peacefulness. [237] Plato referred to music as a moral law, giving soul to the Universe, wings to the mind, flight to the imagination, and charm and gaiety to life and to everything. We could say that the internal and external worlds collide through sound!

Second, ***my thoughts and feelings are part of a Brain-Heart/Mind-Soul Learning Continuum***. [220-222] People are holistic beings with the heart and mind as an integrated biological and complex part of the embodied human system. Within this system, neuroscience findings give us hints of what is possible. There s a brain-heart/mind-soul learning continuum which brings to mind and soul the potential for an existential ate of learning while recognizing the mind/brain human life is still focused in the physical material reality. is coherent whole does *not* insinuate a *linear progression*, but rather a collective working together to ieve growth and expansion, with each element having a different role to play in learning that learning, with each unable to achieve that goal without the other. While they are interdependent, there is a *focus* rms of the *origin of learning* and in *attraction based on the frequency of thought*. For example, iential learning is focused from the viewpoint of the material physical reality. Existential learning is focused from the viewpoint of the soul, that is, from the domain of the higher mental faculties, which have access to an individual's experiential learning AND the larger consciousness field, as well as an understanding of the impact on the emotional and moral development of an individual.

The human journey of experiential learning in the physical material body is one of acting, reacting, then synthesizing, with each new action building on all of the acting-reacting-synthesizing cycles that have come before. [222] We build upon previous learning. Synthesizing is both simplification and explanation, as well as the ability to create a coherent whole from various pieces of things. This ability to decide what information to heed, what can be ignored, and how to organize and communicate what we think is important, is a core competency for everyone alive today.

The existential learning process from the viewpoint of the soul—relating to the development and expansion of thought and action in terms of morality, emotions, and the higher mental faculties—is at the pinnacle of all that has been learned in the physical experience. As we know, energy conserves itself. There is never any waste, merely transitions. While action is no longer a focal point in the existential state, all the patterns created during the acting-reacting-synthesizing experience remain. Nothing of value is wasted.

Recognizing that patterns repeat themselves throughout nature, we can imagine a "being" cycle based on thought versus action, howbeit thought no longer bounded by the limitations imposed by the physical connections of the mind/brain, rather fueled from the viewpoint of the soul. [226] While patterns created during the physically-focused experiential cycle certainly serve as the foundation of the soul's existential state, these are not exact replicas, that is, blow-by-blow details of events and experiences. Thus, they are easily adapted to the thought at hand, and in the existential state can be recombined and expanded to produce a different learning experience. This learning cycle is one of being, reflecting, then planning.

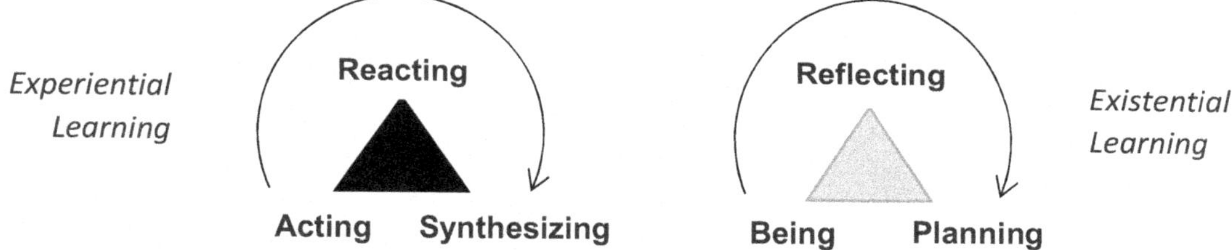

This offers a new direction for the concept of existentialism, no longer tethering the concept to the physical body at the lower end of the Brain-Heart/Mind-Soul Continuum, but rather to the divine nature of the individual human soul at the higher end of that continuum. [323-236] "Divine", derived from the Latin *perfectus* (completed, accomplished, excellent), means perfect, or godlike. From this viewpoint, you exist, you still exist, you always exist as an individuated expression. Yet this concept of existentialism is *in full relationship with concrete human learning experience*, starting with the authentic thinking, acting, feeling individual who has freedom in terms of agency or choice, and the learning and expansion that result from that agency. **It is an existential "experience"** *orchestrated from the viewpoint of the soul*. [225-226]

Mouse: Is this the same for animals? Do animals have this Brain-Heart/Mind-Soul Continuum? Do I have a soul?

Sherpa Owl: I'll share my thoughts on that. First—as no doubt you realize—animals have minds of their own, with emotions. The October 2022 *National Geographic* shared stories about animals showing the complex inner processes of animals. There's Canadian sphynx cat who is very responsive to humans; an octopus who swishes around when her owners come home; a Japanese macaque who stares at his reflection in a mirror, recognizing himself; a sow who's facial movements communicate emotional states; a sheep who can pick out a number he was taught to recognize and exhibits high emotional intelligence with his handler; a raven exhibiting high cognitive functions; and rats who demonstrate empathy.

Sherpa Quercus robur: Let me take that a bit further. Mathematically, a professor named Walter Gomide uses transreal numbers to show that everything that cannot be reduced into logical discourse—and thus is not propositional in its nature—*can* be known through "sentient justification". [177-178] This leads to assessing how we obtain or acquire knowledge from the perspective of sentient agents. Let me break that down a bit. "Sentience" means an intelligence which includes feeling and thinking, and this idea of "sentient agents" is NOT limited to humans since *any living being* endowed with sentience is able to "know". That includes animals. So, while this is complex, it is clear that animals have a mind AND emotions AND the ability to "know", which is reflective of a soul. So, from my viewpoint, the answer is yes, animals have a soul.

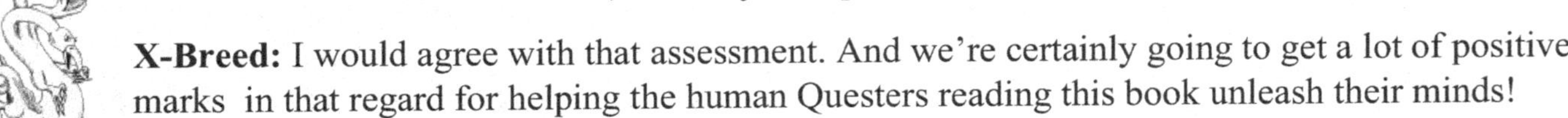

X-Breed: I would agree with that assessment. And we're certainly going to get a lot of positive marks in that regard for helping the human Questers reading this book unleash their minds!

Third, ***my innate ability to plan helps create the direction of the future.*** Planning is forethought—a mental activity focused on achieving a specific goal. Planning, related to forecasting and predicting the future, brings the three parts of time together (past-present-future) and is an interactive part of consciousness. There is always an element of the unseen when we create a plan. This element is our thought, which, while it may be unseen, is *as real as any visible material object*. Remember, as soon as it is made, that plan or thought draws to itself other unseen elements which provide the power to execute the plan or the thought. Unfortunately, when we have fear or expect negative consequences, we are also creating a construction of unseen elements which draw forces similar to that thought. As we have learned through neuroscience, *what we believe leads to what we think leads to our knowledge base, which leads to our actions, which determines outcomes*. [229] You can see how important planning is to our future.

What does this mean to managing self?

As humans we seem to separate things, creating boundaries and naming them. And while setting limits can certainly help us understand things, digging down deeper, as well as sharing these thoughts with others, when we hold onto those as absolutes, then we are bounding OUR OWN thought and growth. Understanding that the soul relates to the development and expansion of thought and action in terms of morality, emotions, and the higher mental faculties, that it is not something separate from our living experience, enables us to seek guidance through that aspect of us, looking at the world and our place in it through a larger framework.

There is power in the human ability to synthesize and plan, and while continuously occurring unconsciously, these are tools that we can consciously engage more fully in achieving our dreams. You ARE the expanded human.

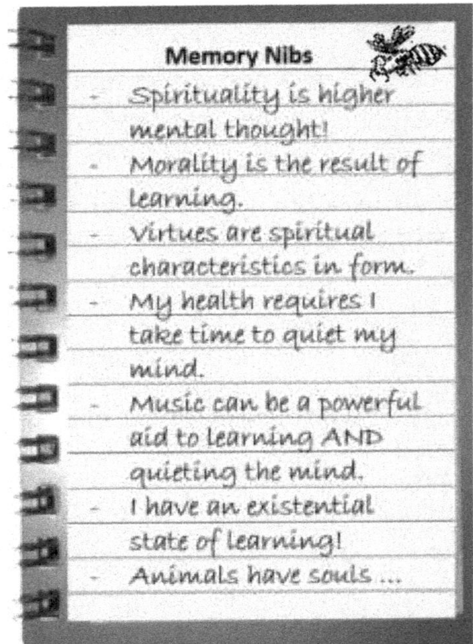

Chapter 11: The Human Gift of Humility [241-263]

Day 11 – Ascent toward the summit; Camp 7

Notes and Activities

Strong. Bold. Competent. Courageous. Adventurous. Humble. There is no conflict among this string of human qualities. These are all complementary descriptive terms representing qualities we admire in our leaders, friends, and avatars. So, what does this concept of "humility" mean exactly? From the viewpoint of the individual, humility is having an accurate view or yourself, neither too high nor too low. This would include knowing your strengths and abilities as well as your weaknesses and limitations, and being honest about them to yourself and others. From the collective viewpoint—the interpersonal level—humility is being other-oriented, focused beyond self. This would include development of empathy, knowing the needs and wants of others, and taking those into consideration in your decisions and actions. [241] As part of an MQI research study to build a deeper understanding of humility, a contrast of opposites was developed. We live in a world of duality, and what some IS NOT helps define what something IS. This is called symbiotic thinking, which is a relationship not from causality, but from existence. For example, if there wasn't a night, would there be a day? Would we have a need for the term "summer" if not also for the term "winter"? Other examples are: good/bad; object/space; matter/antimatter, and so forth. So, we take the time here to share a table of opposites in relation to humility.

	Characteristics Compatible with Humility	Characteristics Counter to Humility
1	Willing to listen; Good listener; Honor and seek truth; Unnecessary to receive rewards for right actions	On broadcast; Talking too much; Voicing/pushing preferences or opinions when not asked; Bragging; Boasting; Using attention-getting tactics; Ostentatious
2	Receptive to difference and new ways of thinking; Having a teachable spirit	Arrogant; "What I have to say is more important"; Inflated view of importance, gifts and abilities; "I'm better than others"; Unteachable
3	Honor others; Serve others; Focus on others in service; Others over self	Selfishly ambitious; Greedy; Lack of service; Serve me; Meet my needs
4	Seek input and perspectives of others; Seek and follow good counsel; Thankful to others for criticism and reproof; Quickness in admitting you are wrong; Repenting wrong actions	Defensive of criticism; Devastated or angered by criticism; Dismiss instruction or correction
5	Honest/open about who you are and areas you need growth; Awareness of faults; Openly address faults; No need to elevate self; Seeing yourself and others as equals; Seeking to build others up; Minimizing other's wrong doings/ shortcomings	Perfectionism; Hide faults; Minimizing own short-comings; Lack of admitting when you are wrong; Defensive; Blame-shifting; Being deceitful by covering up faults and mistakes
6	Gladly submissive and obedient to those in authority	Resisting authority or being disrespectful; Leveling of those in authority; Demeaning; Being sarcastic, hurtful or degrading
7	Gentle and patient; Gratitude; Thankful and grateful to Life; Genuinely glad for others	Scornful; Angry; Contemptuous; Impatient or irritable; Jealous or envious; Lack of compassion
8	Accurate view of your gifts and abilities	Victim complex; Poor me; Focus on lack of gifts and abilities; Complaining; Consumed by what others think of you

9	Possessing close relationships; Recognize value in others; Willingness to ask forgiveness; Talking about others only good or for their good	Not having close relationships; Passing judgment; Using others; Ignoring others; Talking negative about others; Gossiping; Lack of forgiveness
10	Strong, yet flexible	Willful; Stubborn
11	Theocentric; Recognition of being part of larger ecosystem; Realizing higher power	Anthropocentric; Exalts self; or "He is here for me"

Sherpa Prof Owl: Let's do this a bit different. For a year, as people from around the world moved through the Mountain Quest Inn and Retreat Center, the MQI researchers collected data on the state of the world and, specifically, their thoughts about humility as a human attribute. All Questers were provided this material in *Unleashing*, and were asked to take a look at it during the Pre-Quest Workshop. So, let's make this learning session a discussion rather than a lecture, with those who are willing contributing their thoughts.

Unicorn: I agree with the early Greco-Roman culture. Humility is a sign of weakness, and certainly a term that contradicts the strength needed to be leader, so I don't think it goes with those other words, which everyone can agree describe me. I don't need to be humble to know my strengths. I'm perfect.

Quercus robur: Thank you for your thoughts. Indeed, you have many strengths, and there is much value in the diversity of the many attributes represented by this group of Questers! But let us stay focused on this idea of humility. On the surface it would appear that perfection and humility are polar opposites. And, indeed, that does occur, depending on the definition you give to perfection and the personality attitudes that can come along with the idea of perfection. For example, someone who perceives themselves as "perfect" would have no need to listen to anything anyone else has to say, sliding fully into the mode of arrogance. Perhaps we can ask Sherpa Prof Owl to say more about the relationship of learning and humility, ego and arrogance?

Sherpa Prof Owl: Yes, this is an important point. Recall that when we talked about the current environment during our Pre-Quest workshop at MQI, we acknowledge the accelerated mental development driven by the innate human hunger for more—the need to survive and desire to succeed—coupled with a focus on hard competition and economic wealth as an indicator of success. This has led to the expansion of the ego into arrogance, plaguing both individuals and the organizations of which we are a part. Here is what the relationship between humility, ego, and arrogance looks like.

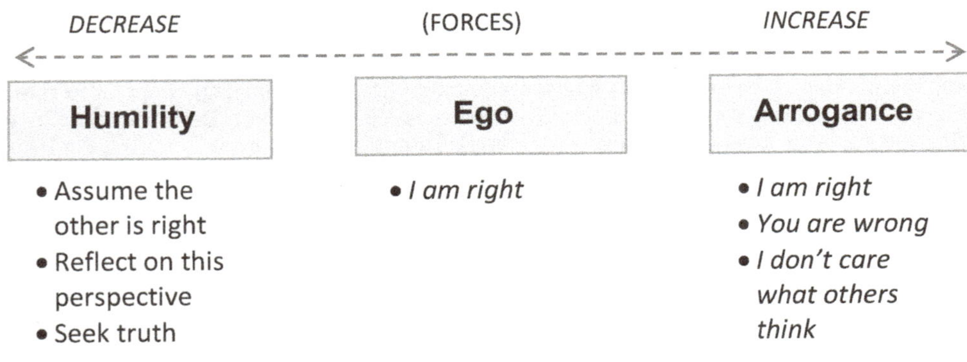

Mouse: Yes, I've seen that in some of my relationships. Ego is self-limiting, and arrogance only drives people away.

Chameleon: Indeed, arrogance can give people the wrong idea or cause them to draw bad conclusions about me. I try to be humble most of the time so I am likable; it makes me more appealing to others.

Sherpa Prof Owl (continuing): There ae many reasons we do things, and we must always ask ourselves if we are doing something for the RIGHT reason. That is worthy of an inner dialogue. Somewhere along this road we seem to have forgotten the nobler parts of humanism, individually and collectively seeking the potential value and goodness of humanity through rational thought. Egotism closes the doors to learning; arrogance builds forces that can lock those doors. When you remedy egotism, the self grows. Since the self is now listening to and considering others' ideas, there is greater opportunity for the bisociation of ideas, and creativity expands. And even a small amount of change can have a large impact on an individual, or humanity at large!

Unicorn: Let's say I accept the definition of humble as knowing my own strengths and abilities and being honest about them to myself and others. So, I could still be humble and still be the best at everything, that is, perfect, which I am.

Quercus robur: Remember that humility includes not only knowing your strengths and abilities, but also your weaknesses and limitations, and being honest about those with yourself and others. Let's go back to this idea of being perfect. From its Greek origins, being "perfect" is being "complete" or "whole". This does not mean being the "best" at something, but rather "whole", a full system. Similarly, the Christian concept of perfection is not as a state of being, but rather as a capacity, trying to be better with negative issues rather than an ability to achieve great things. This is achieving "wholeness", the perfection of the soul. This connection is also made in Islam, with humility the companion of Paradise, and Buddhism, where humility leads to nirvana. Knowing your own limits in terms of the physical, mental, and emotional is part of this perfection. There are always limits. For example, if you abide by ethics and values, you are choosing to be limited. So, in a large sense, limits are a way of exercising humility. Used intelligently, limits can be powerful tools for growth and change.

Dog: I can relate to this idea of humility. Being humble equals being teachable, so it is the key to learning and growth.

Kid: I agree. I think humility is amazing to have. If you didn't have it, then you would never learn or know how to correct some wrong things you have done.

Bees: We believe humility is an essential characteristic for communication and idea-sharing. We've seen this happening as we moved up the mountain. One has to set aside what one is thinking in order to truly listen and engage with others. Being humble helps form a true connection and understanding, and allows selfless conversation to follow. [General agreement]

Sherpa Prof Owl: Excellent insights from all of you. These thoughts reflect the idea of moving beyond one's self, and that idea forms the basis for development of a critical human learning tool.

Quercus robur: Yes, simple, yet profound, the conscious choice of humility is a powerful tool for the discovery of truth. Let me reiterate, that the greatest barriers to learning and change are egotism and arrogance, with egotism saying, "I am right" and arrogance saying, "I am right. You are wrong. And I don't care what you think or say." This shuts the door to learning, ceasing to listen or consider others at all, which is necessary for the discovery of truth as well as growth and expansion. Conversely, as many of you have insightfully shared, humility opens the self to others' thoughts and ideas, providing an opportunity for listening, reflecting, learning and expanding, and the discovery of new truths. Here's the exercise.

X10: The Strength of Humility

STEP 1: To develop humility, first *open your mind* to accept that, by nature, at this point of development human beings have egos and desires, both of which can have strong emotional tags connected to them. It can be quite difficult for an individual to recognize egotism and arrogance in themselves. Remember, the personality, not the self, is often in control, so the individual may or may not be aware of their projection or position. This is potentially true of the individual with whom you are interacting, as well as yourself.

STEP 2: Second, *assume the other is right*. Set aside personal opinions and beliefs for the moment, accept what is being said, the idea or concept, and reflect on this new perspective in the search for truth. While this may prove quite difficult for an individual who is highly dependent on ego and arrogance to survive in what can be a challenging world, almost every individual has someone or something they love more than themselves. Try imagining that this new idea is coming from that someone whom you love, or emerging from something that you love. This simple trick will help increase your ability to engage humility.

STEP 3: Adopting this new idea or concept, *try to prove it is right*, pulling up as many examples as you can and testing the logic of it. If all the examples you can pull up fit this new perspective, then you have discovered a new level of truth. If the examples contradict the concept, then bring in your ideas and test the logic of those. Again, if the examples do not all fit, continue your search for a bigger concept that conveys a higher level of truth. The critical element in this learning approach is giving up your way of thinking so that you can understand thoughts different than your own, discovering new truths. You can compare the various concepts, asking which is more complete.

One issue that may emerge is the inclination for people to think how they feel first, then think about the logical part to determine truth. The "feeling" has already biased their higher conceptual thinking, which may result in it being untrue. It is necessary for us to develop a new sense of self that does not require us to be right in order to feel good about our self.

STEP 4: Once we come to a conclusion, we need to take action. It is time to affirm our incorrectness as appropriate to those with whom we have potentially lacked humility, and to *show gratitude for them sharing their thoughts with us*. Note that the expression of appreciation and gratitude reduces forces. It is not enough to say that you were wrong, *nor is that an important issue*. What *is* important is to acknowledge that someone else is right, and that you are appreciative of learning from them.

STEP 5: Finally, *ensure that your motive for adopting humility is your search for truth*. Motive eventually comes out, and the wrong motive will defeat the purpose at hand. In this search for truth, you are using mental discipline to develop greater wisdom. It is difficult to overcome the urge to "look good" and to be "righter" than others. But remember, when we are "full" there is no room for new thought. When choosing humility as part of our learning journey, we discover that it is not about being right, rather it is about the continuous search for a higher truth.

GROUNDING

Energy is coursing through the bodies and minds of humans and critters alike. And much like electricity, the grounding of self is an important and necessary part of life. While there are many ways to do this, humility requires grounding ***of*** self rather than grounding ***in*** self. Conceptually, grounding can be considered (1) providing a foundation; teaching the basics (family, culture, environment, religion); (2) fixing something in or on something else (material goods, people, beliefs and faith, ideas); or (3) connecting with the ground (Earth) (nature, part of larger ecosystem). Grounding in or on something else often involves possessions, level of education, successes, wealth, control and power, which often leads to an expanded ego and arrogance. Conversely, grounding through positive relationships with others and nature, and living as part of the larger ecosystem, supports development of morality, service, joy, and love and contributes to the greater good. The choice is yours. *Consider:* What is the most important in your life?

Eagle: There is something to this idea of humility. I can clearly see how it would work well with others who have a level of intelligence and knowledge in an area I am unfamiliar with, but I find it hard to consider that creatures like the sloth, snail and nematode actually have the ability to contribute much other than playing a role in the food chain. Still, since the lion stopped the sloth's fall, the sloth is keeping his eyes open for longer periods of time and asking questions in our sessions. I wonder what might be learned from that?

Sloth: Somehow, I don't seem quite as tired as before, and my head seems to be working a bit more actively as well. And while I may not contribute a whole lot to the conversations we are having, even with my eyes closed I am listening and doing some thinking about what is being said. I can only attribute that to my fall being somewhat of a "wakeup" call to other possibilities in life, although I still love my mid-day nap when I can get it.

Ants: We think humility is about respect. Respect for others is really important! We've learned that the human being is equipped with a very powerful organ, the brain/ mind, which has the capability of creating wonder. The least one can do is respect it!

Quercus robur: Wisely said. There is a strong relationship between humility and respect, whether that respect is based on the individual's brain, experiences, lifestyles, nationality, culture, beliefs, knowledge, or many other aspects of what it is to be human. This is the idea of honoring others.

Eagle: There appears to be a need for balance here. Every structure we see in the Universe is a result of the balancing between opposing forces of Nature. I recall an Australian educator and businessman saying that the natural balance that exists in Nature is something rarely achieved in human systems. So, how do we find that balance with reference to humility? Anyone have any thoughts?

X-Breed: Humility is certainly important to the extent that we're open to new ideas, but not to the point of being a doormat.

Unicorn: Hear, hear! It can be a double-edged sword which others take advantage of if you aren't careful!

X-Breed: I was referencing a willingness to defend what and whom you love. And also, the willingness to speak your mind openly and honestly. If not, you are betraying your life.

Jackal: Pride in work and personal accomplishment contribute a lot to conversations and allow people to get to know another and share interests. Pride also demonstrates passion, which gives human connection. Bragging and hubris are different than a well-deserved feeling of accomplishment, and they belittle yourself as well.

Triceratops: I regret being too humble in the past. I've done a lot of good things, but if you never let people know your abilities and knowledge and accomplishments, you miss opportunities not only to use those, but to build relationships based on that respect we were talking about. I need to reflect on this a bit more.

Quercus robur: Thank you all for your forthrightness and willingness to share. There is a lot to think about in this exchange. Let me wrap up tonight by sharing with you that in the MQI research study—engaging individuals who are in 14 different countries and with a heritage from 35 different countries—agreed that humility is valuable when engaging others. And as one said, "Being humble is an awareness of the greatness in which we are existing and that Life is One." Tomorrow, we reach the summit. Sleep well

SECTION IV:

The Summit—The Quest and Beyond

Living is Learning; Learning is Living

They made it! For some of the critters this seemed miraculous. Yet, they had just kept learning, thought after thought, step after step, as they climbed toward the summit. And they made it! As they stood at the top of the mountain, a soft breeze accenting the fresh air filling their lungs, there was a sense of freedom. It wasn't so much a feeling smarter, but rather the growth of possibilities. And as they exchanged their thoughts—without speaking a word—they felt connected. And slowly the thought emerged within, that THIS WAS JUST THE BEGINNING. There is so much more to learn! The authors join the conversation.

Yak: Do you all remember the Memory Nib about *presencing*—paying attention to what is happening now?
Kid: I get it, but I'm really having to think about all the old and well-used tendencies of my self to not plan for what's coming.
Dog: It's definitely helped me to listen more to others, like the Sherpas, to put my best paws forward as they say.
Yak: You know it Dog!

X-Breed: Hey, Uni, this quest isn't exactly what we imagined. Somehow, after the workshop, I kept hearing "arrogance is not a substitute for experience."
Unicorn: I get it. You're telling me my self has a lot to relearn also. Reflection definitely helps and I love the Xercises. I've always thought I was perfect until I did that first Xercise *Who is My Self?*
X-Breed: I hear you. We have another Xercise coming up called *Expanding the Quantum Self* for really getting a handle on our self. Might prove valuable for both of us.

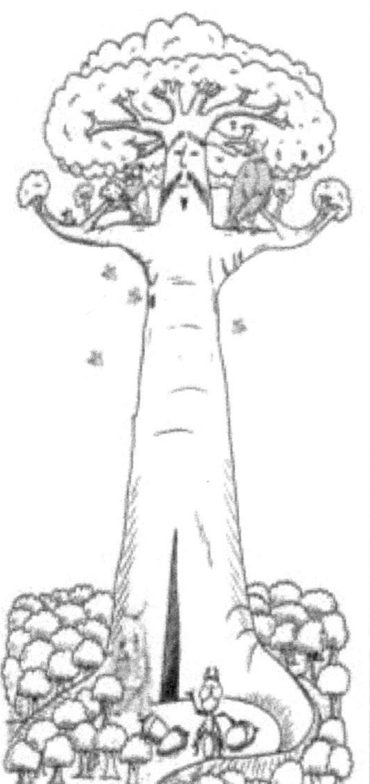

Eagle: Reflecting raises the importance of this journey! The world is changing, and each and every mind is important.
Prof Owl: I'm so appreciative of being asked to participate, and I've learned from all of you.
Beneficial Insect [nodding]: Thanks for allowing me to help.
Quercus robur: And so, we reach the top of this ladder and step onto the first rung of another, reaching for the stars with humanity in tow. What a wonderful beginning!

Jane: Dog, you were energetic and enthusiastic on the quest climb. Were you challenged?
Dog: Dog-gone it! You better believe it was hard. I had to tell my self to adjust my normal behaviors and be careful and serious. Totally new behaviors for me.
Jane: How useful was that for you?
Dog: Woof! I'm a changed dog. My days chasin' after other critters are over!

Gibbon: Tri, I've had so much fun swinging around on your horns!
Triceratops: Actually, you were swinging around on my nose, too, but since that's hard to scratch, it worked out pretty well. So, keep swinging. Ha, ha!

Feline: Do you remember a few days ago you asked me about affective attunement?
Mouse: Yes, that was an interesting session.
Feline: After some reflection, we just may be attuned, and, more recently, my feelings towards you seem to be changing, but don't push it!

Mark: I love spending time with you Jackal, we always have a great time. I hope you enjoy our conversations?
Jackal: Thanks Mark, before we set off I was a little skeptical of how you would contribute to our quest, but you have been great fun to be around, and talented too! I thought I needed to be in control; however, I quickly realized that we needed to spread the responsibilities and tasks for us to be successful.
Mark: That is so true, we often try to tackle everything ourselves. Some say if you want something done, it is better to do it yourself. I say if you want something done, teach someone how to do it.
Jackal: I love that, I will try that next time.

Robert: Lion, why did you change your mind about coming on this quest up Mountain Quest II?
Lion: I peaked at the big **Unleashing** book and couldn't put it down. Kept reflecting on what was said about humility, and it helped me understand that I could be a stronger leader if I reoriented my self to value true regard for the needs of others.
Robert: Why did you help the sloth?
Lion: Because I could. I felt someone might need me and wanted to be there for anyone who did, so I stayed near at hand. He needed me.

Arthur: Gibbon, thank you for demonstrating just how important humor and socializing ideas are for success!
Gibbon: My pleasure. I am pleased that we experienced how people are under stress and that we can calm the environment by sharing a few smiles together.
Chameleon: I thought I was the change champion of the critters, but I realized I just change without thinking about it, whereas you, Gibbon, use fun as a strategic tool for building relationships, collaboration and learning!
Arthur: This shows just how much we have to offer each other when combining our strengths to resolve challenges together.

Alex: Have you noticed a swishing in your head? It's like a whole new space has opened up, just like when I was trying to learn Calculus over two years and it finally took hold!
Quercus robur: It's the North Wind heading South and the Sun rising East and setting West. It's living.

1st Bee: Hive Bees, I'm not surprised they asked us to keep track of the *Memory Nibs.* Anyone have a favorite nib you're passionate about?
2nd Bee: with all the misinformation being passed around these days, one that resonates with my self is *Critical thinking supports effective thinking.*
3rd Bee: I have to remind my self of that every day now!
Other Bees: Don't we all!

Whale: The next quest must be diving into the depths of the ocean! There is so much to learn in the flowing inner realms of Earth, so much to share with all of you.
Snail (always listening): Slug! When does that quest start?

A Special Gift for YOU

The OrgZoo Questers would like to offer a special gift for all the readers who have joined them in this exciting quest, sharing all they have learned with YOU. So, open your mind, and they will each "connect", some via their Self Phone and others directly mind-to-mind, a self capability many seem to have opened to during this quest! Joining us are the Eagle, Owl, Quercus robur, Ants, Bees, Chameleon, Dog, Feline, Gibbon, Jackal, Kid, Lion, Mouse, Sloth, Triceratops, Unicorn, X-Breed, Yak, and Whale. We learn from them all! Note that some of the OrgZoo characters (who have not yet learned to use their brain directly for these purposes) need to hook up using an old-fashioned hard line. Now, close your eyes and just stay calm for a few minutes as they navigate the cortical columns in your neocortex, hooking up with the appropriate reference frames where you have already been developing a relationship with their thoughts.

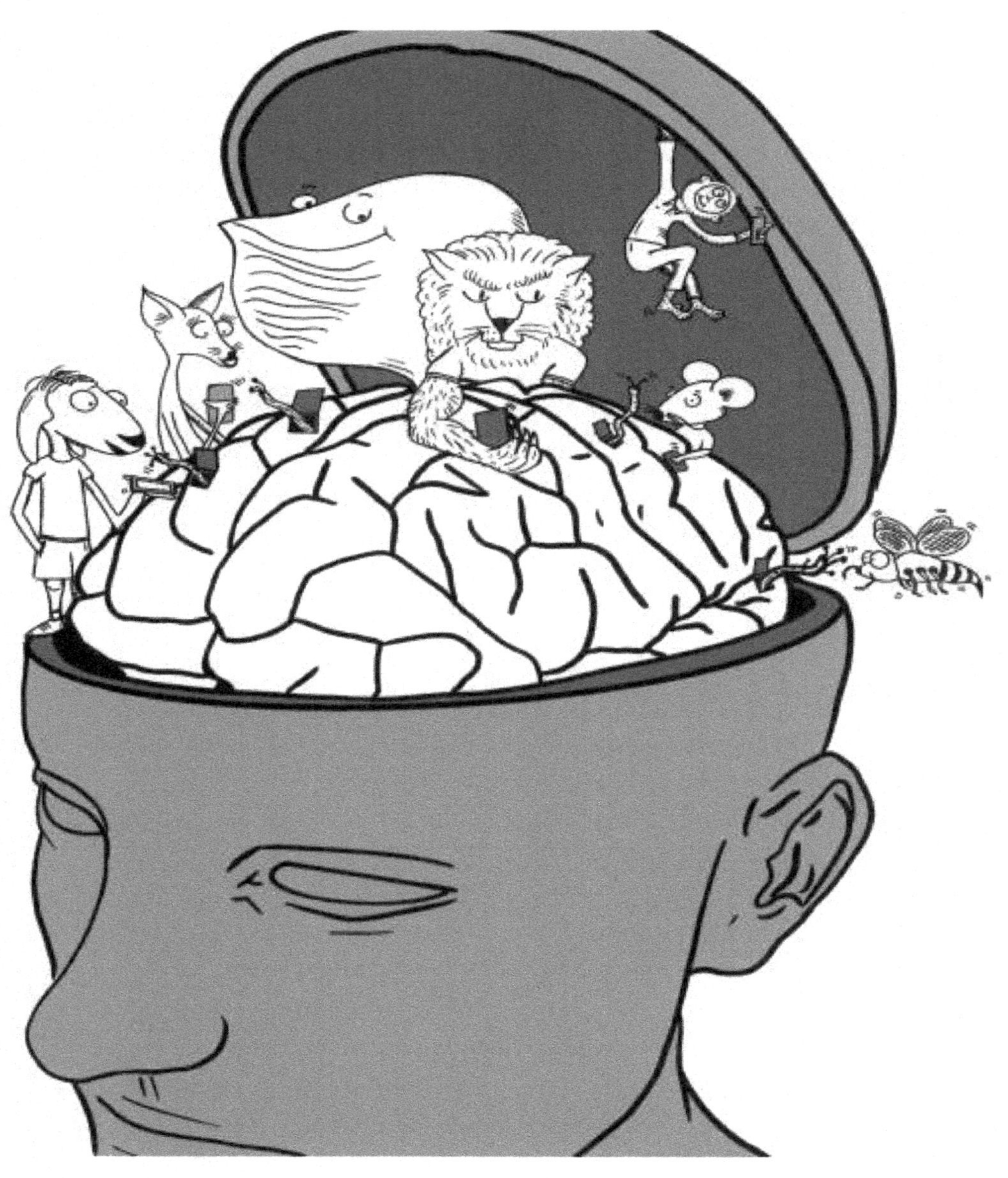

Choosing Consciously

Where do we go from here? When a chaotic world seems to be squeezing air out of your lungs, how many times have you asked yourself that question? We know that we are complex adaptive systems, and that a CAS cannot exist in stasis; you must either learn or decline. The choice is ours. *To get the most out of life—to be the best version of the possible you—is an outcome of the conscious choices we make.*

Choosing consciously begins with awareness, awareness of self, awareness of others, and awareness of the situation at hand. Awareness is the beginning of each development cycle we undertake: Awareness, Attitude, Ability and Action. Awareness (informing your mind through listening, questioning and critical analysis) leads to consciously choosing an attitude (understanding how we feel about this situation and why we feel that way). These two steps combined highlight what abilities we need to apply or develop to enable us to take constructive actions. When we consciously choose what to do at each stage of these developmental iterations, we accelerate our progress and increase our adaptability.

Choosing consciously is being adaptable. Without consciously choosing what we do next (informed by what we, and observed others, have done previously), we fall into the subconscious patterns of our past. Sometimes reacting based on past patterns is OK. It can keep us safe and does not challenge our comfort zone. This is fine, if the world remains the same. However, the stark reality is the world does NOT remain the same. The events of each Earth rotation changes perceptions and realities. Our world becomes different, and to better engage with this constantly shifting environment, we benefit from being adaptable.

This is a lesson we can learn from the critters who joined our quest up the mountain, as well as the critters who chose to remain in their known day-to-day patterns. Sometimes we are defined by what we choose to do and with whom. However, we can also be defined (positively or negatively in the eyes of others) by what we choose *not* to do and with whom we choose to *not* interact. Some people say they do not get involved in politics, perhaps not realizing this is in fact a political statement. Every action we take is part of defining who we become, including the decision *not to act.* Consciously choosing enables you to make more informed choices and to understand yourself and others at a deeper and broader level. If you are not consciously choosing, your subconscious mind will make the decision on your behalf, and this may not align with your conscious plans for your best future.

Choosing consciously is thinking, voicing and acting out who we choose to be (or desire to become). Remember, YOU are a verb, not a noun. We live in a state of becoming. We are defined by the sum of our perceptions and experiences and how we consciously choose to put these into action. That is, we are constantly becoming *what* we choose to think about and act on (and what we do not), *how* we choose to do this and *why*. Of course, not all of our actions turn out the way we imagined they would. That is life! Yet every choice, whether with results as predicted or otherwise, makes a difference in *who we are becoming.*

Our thoughts count. Thoughts are influenced by both conscious and unconscious patterns. Understanding your thoughts and the elements and experiences that influence them is critical for success. Thoughts are invisible and yet powerful. Everything that exists in our reality was a thought first.

Our words count. Words, and your perception of what they mean, are important, as is the way we consciously choose to act on them. The world continues to revolve and evolve whether you make conscious choices or not. You can choose to simply roll with what happens, or you can make conscious choices to influence who you become, how you interact with the world, and therefore what you can achieve in life, and these are communicated through your words as well as your actions.

Our actions count. Actions are visible, and yet prone to interpretation by others. People make judgements about what people do, despite not fully understanding their reasons for doing this. We perceive the reasons for other's actions based on our own experiences, rather than through the lens of the person acting. *Even the best of actions can be misunderstood.* This is why consciously choosing requires your full attention as a living complex adaptive system, reflecting the inner you in your interactions with your environment in real time.

Arthur: We are all prone to feeling relaxed in our own "Comfort Zone", which provides a sense of security and familiarity and reduces the anxiety of not knowing what will happen next. Most people develop such behavioral patterns to avoid stepping outside this comfort zone and feel safe. We find all sorts of excuses about why we do what we do (and don't do).

Since all change begins with awareness, a good habit is to make a conscious effort to challenge your initial feelings about anything that you seem to be automatically rejecting, or preferring. Such reactions are generally made in the subconscious mind, and this is not necessarily the best way to decide how to act. Since consciously choosing to do this, I have discovered many things that I would never have otherwise experienced!

One example is when I was in a conversation with a business leader in Poland who referred to me as creative. My reply was, "That is ludicrous, I am a scientist and a project manager. I work entirely on the known facts". His answer was insightful and changed the way I perceived myself. He simply stated, "You do not know yourself – you ARE creative in many ways. I see it in what you do every day". After reflecting deeply on what I do and why, I realised that there were many things that we just do and never think about much in the doing, but just taking those as "just the way it is" or "just the way we are". Things like how we interact with people, how we use metaphors when influencing people and engaging stakeholders, the way we solve problems, and more.

Reflecting deeper after this conversation, I realised that I had always been attracted to complex challenges and enjoyed creating novel solutions. And I recognized that I'd done this throughout my career, and that it was a significant part of my success. This new SELF-insight triggered a new self-perception that triggered many proactive creative activities, such as discovering the critters and their characteristics, writing the OrgZoo book and associated games, and building an OrgZoo Ambassador network, international communities to collaborate around soft skills. *Choosing consciously is not a remedial approach.* You are not broken and in need of repair. Choosing consciously brings you to a higher level of awareness, enabling you to become the best form of you, as you continuously find new ways to learn and grow.

As together with the critters we the authors stand at the summit of this journey, and begin another, we thank you for joining us and the OrgZoo critters on their quest. You will appreciate that the process of creating the Field Guide with these critters helped *us* unleash our minds from a more academic style of writing, and, well, it was just plain fun! These critters are quite wonderful characters and rather friendly chaps at that.

We trust that you were also unleashed in some useful way and now find your self with valuable new insights. And, while we still have your attention, here are three additional Xercises to support your continuing journey of expansion and unleashing: *Being Adaptive*, *Expanding the Quantum Self*, and a game focused on *Applying What We Have Learned.*

X11: Being Adaptive

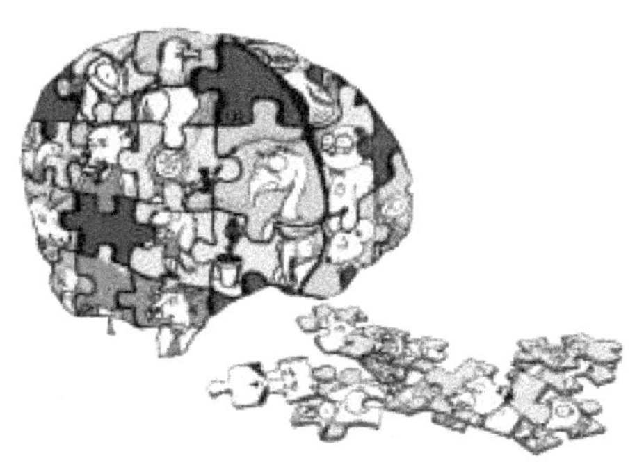

Choosing consciously works best when we apply all three aspects of capability—knowing, doing, and being—in the context of our situation. Knowing comes from reflecting on the knowledge available and determining what is most relevant. Doing comes from drawing on our skills and talents to determine how to act on that knowledge. Being is how we choose to interact with others in the context and spirit of the moment consistent with our inner self—beliefs, values, and virtues. Each of these three elements of capability can change in the moment in emergent and complex situations, which is why being adaptable is critical.

Since being adaptable is a significant part of sustained success and ongoing development, it draws on sensemaking, emotional intelligence, empathy, and critical thinking to make better judgements about which behaviors are best matched to your current situation, and when to shift from one situation to another. Inviting the critters into this Xercise, the table below provides an example how each critter's characteristics/behaviors would be helpful and unhelpful. Interestingly, although the behavior is the same, the outcome is different in different situations. In other words, we choose when specific behaviors are effective and when they are not, adapting to the situation at hand.

STEP 1: Browse each critter and think about an example when you displayed this behavior when it was helpful, and another example when it was unhelpful.

STEP 2: Reflect on the outcomes from each and why this was the case.

STEP 3: Practice each of these behaviors using safe conversations to familiarize yourself with them, aiming to build your confidence and competency to consciously choose to apply them when they are necessary to apply in your everyday life.

Critter (Behavior)	**Matched behavior (Helpful for context)**	**Mismatched behavior (Not helpful for context)**
Ant	Getting things done when strong task orientation is required	Focus on task completion when creativity and alternative thinking required.
Bee	Inclusive interactions and communication in collaborative team activities.	Aggressively defending the team despite them doing unhelpful activities.
Chameleon	Fitting into the current status quo to go with the flow.	Not challenging others when alternative options are better.
Dog	Supportive of the team with absolute loyalty.	Blindly following leadership into a crisis without critical thinking.
Eagle	Inspiring others as a role model for decisive action.	Acting too swiftly when a considered decision may have been more constructive.
Feline	Ensuring productivity among team members.	Playing mind games to reinforce their own position in the hierarchy.
Gibbon	Leveraging humor to reduce tensions.	Not focusing on the key issues in a crisis.
Hyena	Networking with colleagues to ambush challenges through shared insights and resources.	Politicking to call on return favors that may compromise others.
Insect beneficial	Providing insights and experiences to accelerate resolutions.	Advocating for a particular option when a range of alternatives may work.
Insect pestiferous	Provide a variety of ideas from external sources.	Self-interested actions being advocated which may not be best options.
Jackal	Social care and emotional support for the group members.	Taking advantage of privileged position, often leveraging deferred power.
Kid	New creative (naïve) insights that stimulate creativity and innovation.	Offering naïve ideas that cause distractions for resolving a crisis.
Lion	Controlling processes and resources to ensure safety and productivity.	Leveraging position to force others to follow directives against their will.

Mouse	Engaging in productive work when focused and well managed.	Engaging in non-productive or even destructive activities when bored.
Nermatode	Indulging oneself to relax when on holidays, so you return to work completely recharged and ready.	Almost any work situation—taking benefit without giving back is unforgivable and bound to destroy relationships and damage culture
Owl	Mentoring others to become adaptable and knowledgeable.	Rarely, perhaps being too open and not pushing for mentees to act more quickly in important developmental needs.
Piranha	Aggressively tearing a problem apart to remove the issue.	Engaging in damaging gossip to undermine people.
Quercus robur	Philanthropy and giving are almost always beneficial when well directed towards deserving recipients.	Rarely, perhaps, too soft on some who may be taking benefit, but not fully deserving of the support.
Rattlesnake	Highlighting they are challenged by your presence by making a lot of noise about an issue (without striking).	Making a lot of noise about minor challenges.
Sloth	Waiting to allow things to settle for themselves rather than blowing them up into big issues.	Not acting fast enough to correct situations before they become a bigger issue.
Triceratops	Stubbornly refusing to change things that should be maintained as is (for good reason).	Stubbornly refusing to change to better approaches with good evidence to support the benefits.
Unicorn	Creatively offering very different options without restraint.	Offering alternative ideas that are totally disconnected from reality of the moment that are simply impractical.
Vulture	Finding weaknesses in the plan or system and taking out the weaker options before they become an issue.	Criticizing problems after they have happened (when perhaps they could have highlighted the risks earlier, in time for preventable actions).
Whale	Brings intellectual insights into the conversations, especially around specialist areas like technology, knowledge and science.	Sometimes too smart for their own good and beach themselves because of their certainty of being right.
X-Breed	Provide a range of perspectives into the conversation based on prior experiences and learning and recommend which they believe is best and why.	Overconfidence based on their multitalented capabilities and knowledge can lead them to make a limited decision that does not consider wider or longer-term implications.
Yak	Some of the greatest inventions come from errors, for example, penicillin. Enthusiasm to act and see what happens can generate new insights.	Actions without appropriate reflection can cause a lot of unpredicted collateral damage, both in the short term and longer term.
Snail (Hidden background behaviors)	Getting in to do the basic routine activities in background, without the need to be noticed or provided with kudos.	Background routines can become quite outdated and inefficient without regular review. Limited attention to such activities can become poor habit.

X12: Expanding the Quantum Self

Introduction: *We can identify self as a critical node—and learning as a primary catalyst—for state change (quantum change) in human and living systems.* [5] When we reflect on "mountain" and "summit" experiences in our lives, we become more aware of the potential for unleashing our mind. This is wisdom that increasingly includes the breakthrough knowledge that our mind expands by increasing our intention to manage our self. Engaging our human capacity, we are empowered as we move forward and upward, sensing that we are lifted up. As we ascend, we see that the summits on our journey reveal new vistas—our self envisions new potential.

To that end, this Xercise is designed to enable you to learn to access and use learning capabilities in more dynamic and productive ways. As you use it, following the simple basic process steps, the steps will be encoded in your self. Essentially, this process will help you intentionally manage and facilitate your learning and thinking patterns and processes. As this is achieved, your self, as a core node in your human system, will adapt and grow from enriched learning, think more effectively, and participate more fully in intelligent and purposeful living. In short, your self will expand and unleash your human mind.

Setting Up Your Quantum Self Xercise

This Xercise is rather straightforward and, as such, it can be easily and frequently used. The process will prove fascinating as you set it up and experience it. Your result can be transformative.

STEP 1: First, find a quiet place and briefly review the Quantum Self Model at the right. For definitions of the nodes, please refer to the Glossary (Resource E). As you become increasingly familiar with the nodes over time, you will become more knowledgeable and fascinated with their individual functions and interactions.

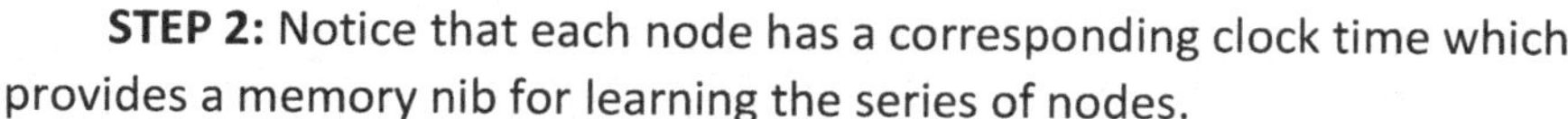

STEP 2: Notice that each node has a corresponding clock time which provides a memory nib for learning the series of nodes.

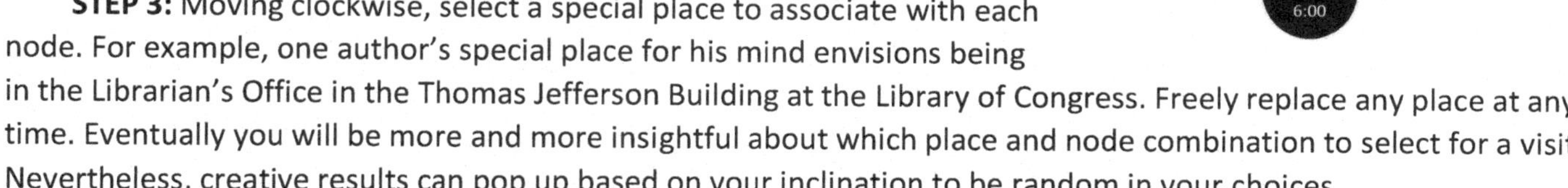

STEP 3: Moving clockwise, select a special place to associate with each node. For example, one author's special place for his mind envisions being in the Librarian's Office in the Thomas Jefferson Building at the Library of Congress. Freely replace any place at any time. Eventually you will be more and more insightful about which place and node combination to select for a visit. Nevertheless, creative results can pop up based on your inclination to be random in your choices.

STEP 4: Write each place down by its respective node on a separate sheet of paper where you can keep notes.

The Process for Using Your Quantum Self Xercise

STEP 1: Relax where you will not be interrupted.

STEP 2: To begin, identify an important question for which you seek an answer. This could be about yourself, or about learning, creativity, innovation, planning, decisions, improvements, relationships, challenges, formal education, employment, wellness, or anything of significance to you. Write it down on the paper with your special places. Creating the question will help you to work with it in both your conscious and subconscious levels.

STEP 3: Now, choose a place and visit there, taking your question with you. Your question is the focus of your visit. The nodes and their related spaces will share an increasing bond to help you learn how the nodes work and to have more vivid sensory relationships with them. The spaces will also trigger unique, emotional, and insightful perspectives for pondering your question.

STEP 4: When you use this Xercise, you may choose to visit more than one place. As you visit your special places, reflect on how you feel about being there. It is invaluable to be aware of your emotions while learning. Also, each time you visit a special place, see if you easily notice something for the first time. The clearer your picture of a place is in your mind, the faster and more effectively you will be able to process it.

STEP 5: Now, ponder the thoughts that come to you about your question as you visit your special place. Be careful, taking notes as prompted. A small matter may have major "quantum value".

STEP 6: While visiting a special place, you may wish to imagine what someone you regard would ask you about your question. If you interact, be careful to remain focused on your question.

Future Consideration

During this decade, it is anticipated that technical advances beyond virtual reality into mixed reality, which includes augmented reality (AR) will enable experiencing this Xercise with increasingly powerful results. When that occurs our choices for special places will be enhanced by computer hardware and software to provide a much faster and more vivid sense of reality with heightened sensory perception and increased interaction. This will likely include real-time messaging and collaboration with others in special shared spaces, reminding you of the OrgZoo Self Phone.

X13: Applying What We Have Learned

Learning is the creation of knowledge, and knowledge is the capacity to take effective action. We also now understand that everything in our physical reality was a thought first, and that our intention and attention create our reality. This begs the question: How have our thoughts changed and our actions shifted as we have progressed through the unleashing process? That question is at the core of this Xercise. First, print out pages 65 and 66 back-to-back and pages 67 and 68 back-to-back. Following the cutting lines on pages 63 and 65, cut out the 28 boxes ("cards"). Each card will present one neuroscience finding from the ICALS research. Bring all the cards together into a deck with the word "Unleashing" showing and the neuroscience finding hidden.

Playing as a group sit in a circle and decide who will be the first "Questioner". The person to the left of the Questioner is Player 1, who draws a card and reads it to the group. Now, it is the role of the Questioner to decide which of the following questions Player 1 will answer:

(1) What does this mean?
(2) How does this change the way I think about my SELF?
(3) How can this change the way I manage my SELF?
(4) How does this affect the way I interact with others?
(5) How can knowing this improve my ability to co-create my reality?
(6) What is one example of how this can be applied in your everyday living experience?

Following Player 1's response, moving around the circle to the left, other players have the opportunity to agree that Player 1's response showed an understanding of the ICALS finding OR expand on that understanding, providing additional response to the question. Player 1 receives one point for answering the question and one point for every player that agrees. However, when a player chooses to expand on what has been said or redirect the response, this expansion/redirection is now the focus and this player receives one point for expanding and one point for each player that agrees as play continues moving around the circle (the initial player responding does not receive any additional points once an expansion/redirection has occurred). As this continues, when a player provides an additional response, this expansion/redirection is now the focus and this player receives one point for expanding and one point for each layer that agrees, and so forth, following this cycle as long as another player has something to expand on or redirect the response. Note that *only the players either answering the question or participating in the expansion process gain points.* An "agree"—which in essence is a "pass"—provides additional points to the previous player who has responded or expanded on the response, but not to the player who agrees. When all players "agree" and the circle has returned to the last expander, the initial Player 1 becomes the Questioner and Player 2 begins the response.

NOTE: While everyone can expand on a response, ONLY THE INITIAL QUESTIONER CAN "DISAGREE" WITH A RESPONSE OR EXPANSION OF THE RESPONSE, and when this occurs it is up to the Questioner to convince the players why this response or expansion on response does not work. When a majority of the players agree with the Questioner, the Questioner automatically gains five points, and the cycle begins afresh with a new Questioner, a new Player 1, and a new question.

A game consists of one full cycle, that is, with everyone having the opportunity to be a Questioner. Since when there is a large group this could become quite time consuming, it is suggested that the larger group be broken down in to 6-8 people for optimal participation and learning.

Playing as an individual, choose a card from the deck and see if you can answer all 6 questions. *Hint:* Following the ICALS finding on the card, the page #s where the answers appear in the full *Unleashing* book is in brackets. Points are accumulated by each question answered. When a question cannot be answered, use the *Unleashing* book to help discover the answer before going on to another card.

The best mental exercise to slow aging is new learning and doing things you've never done before. [161]	The mind/brain creates an internal representation of the world. [3]
The neocortex constantly tries to predict the next experience. [38, 229]	An enriched environment can produce a personal internal reflective world of imagination and creativity. [161]
A model of the self comes mostly from the unconscious. [121]	Volleying between the conscious and the unconscious increases creativity. [106, 111]
Emotions influence all incoming information. [9, 166, 270]	The brain actually needs to seek out an affectively attuned other for learning. [191]
What we believe leads to what we think leads to our knowledge base, which leads to our actions, which determines outcomes. [115, 229]	Genes are not destiny. The environment can change the actions of genes via non-expression. [114]
Volition is necessary for benefit. Forced exercise does not promote neurogenesis. [158]	Meditation and other mental exercises can change feelings, attitudes and mindsets. [103]
Physical activity increases the number (and health) of neurons. [159]	Memory stores invariant forms (used to predict the future). [130]

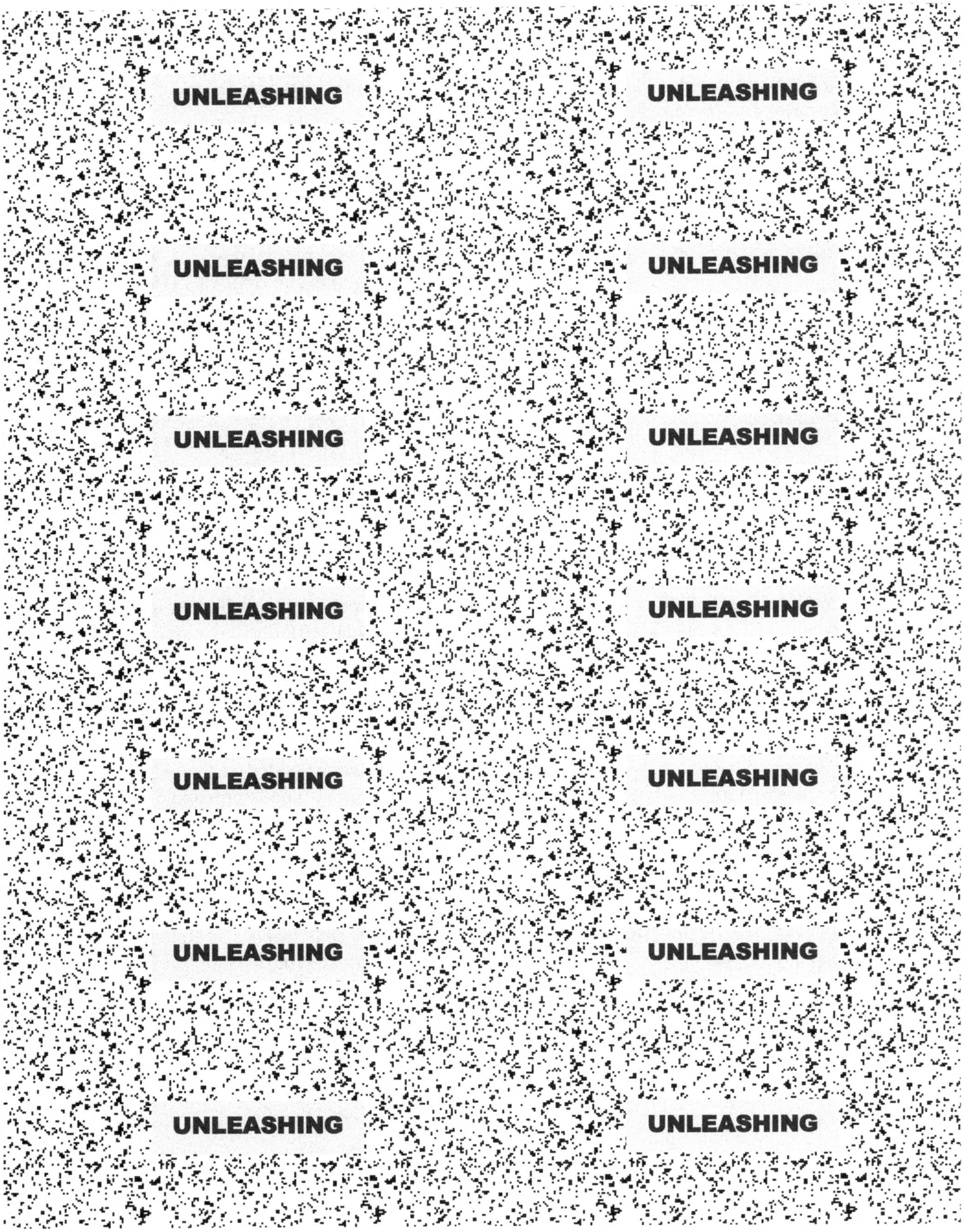

UNLEASHING

UNLEASHING

UNLEASHING

UNLEASHING

UNLEASHING

UNLEASHING

UNLEASHING

UNLEASHING

UNLEASHING

UNLEASHING

UNLEASHING

UNLEASHING

UNLEASHING

UNLEASHING

Repetition increases memory recall. [146]	Memory recall is improved through temporal sequencies of associated patterns, that is, stories and songs. [132, 273]
Neurons create the same pattern when we see some action being taken as when we do it. [195]	Mirror neurons facilitate neural resonance between observed actions and executing actions. [196]
Thoughts change the structure of the brain, and brain structure influences the creation of new thoughts. [88]	Plasticity is a result of the connections between neural patterns in the mind and the physical world—what we think and believe impacts our physical bodies. [7]
An enriched environment increases the formation and survival of new neurons. [16]	Affective attunement contributes to the evolution and sculpting of the brain. [192]
Physical mechanisms have developed in our brain to enable us to learn through social interactions. [189, 279]	The unconscious can influence our thoughts and emotions without our awareness. [123, 176]
There is an optimum level of stress for learning (the inverted "U"). [148]	Voluntary learning promotes Theta waves that correlate with little or no stress and positive feedback. [84]
The unconscious brain is always processing. [11, 122, 145, 216]	The unconscious never lies. [125, 308]

UNLEASHING UNLEASHING

UNLEASHING UNLEASHING

UNLEASHING UNLEASHING

UNLEASHING UNLEASHING

UNLEASHING UNLEASHING

UNLEASHING UNLEASHING

UNLEASHING UNLEASHING

Afterword

We live in an age of our world that is bursting with unprecedented need and opportunity. Now, we see more clearly than millennia of our human species have imagined what we're capable of reaching. In our resolve of hope, we know we will find new capacity for living on this planet that transforms the human endeavor. All around us, in every aspect of life, we will overcome that which shatters and diminishes. To do this, we must learn more and better about why we learn—we must learn deeply and powerfully how we can learn beyond our mere survival and through the brilliance of human experience never more to be wanting.

What we boldly undertake brings fresh insight and knowhow for learning in all ways of living. We see that within each of us there is more than a measure of untapped capacity. We look forward with new perception, knowing that the human brain has a reservoir of potential waiting to be called upon. As this century continues to unfold, what you and all of us learn heralds a new Golden Age.

Where there is need, human caring will achieve unbelievable lifting. Where there is devastation, innovation will rebuild. Where there is immeasurable struggle for governance and security, the voice of kindred spirits will roll in the reverberation of the celebration of freedom. Where there is pain and suffering, our sciences will find resolution. Where there is longing, there will be confidence. Where there is evil and corruption there will be transparency and fortitude to turn the tide. With new hope, new vision, new resolve, we will together achieve a quantum breakthrough in learning and creating. To that end, adventure with us in expanding self and unleashing the human mind.

Can't believe they all made it … maybe there's something to this unleashing!!

Resource A: Guiding Principles for Using the OrgZoo Characters in Facilitation

People have often asked what the rules are for facilitating OrgZoo activities. However, rules are not the optimal way to achieve success as they assume control and are too rigid. Rules can apply where a situation is absolute, inanimate and static. In contrast, life, relationships, personal and professional development, and culture are not absolutes. They are complex, subjective and unpredictable, making it inappropriate to attempt to bind them into a set of rules.

Social interactions are deeply influenced by the diversity of those involved, and their perceptions. Opening up conversations to constructively explore difference brings creative learning and builds relationships and trust. Closing out or criticizing others, often because of ill-informed subconscious bias, destroys trust and relationships. An effective facilitator needs to be critically aware of the dynamics in the room. It is the facilitator's role to engage participants in sharing and to ensure their psychological safety.

The guiding principles remind us of the wider contexts in which people interact and how changeable this is. These principles can apply to any facilitation around social learning and are the foundation of safe OrgZoo facilitation:

1. Behavioral adaptability is fundamental for sustained success.
2. Proactive, conscious choice of behavior is better than subconscious reaction.
3. Inclusive Conversations That Matter are critical for mutual understanding.
4. Culture is defined by the more influential animals in your Zoo and their relationships.
5. All OrgZoo characters can be found in most social contexts; some may be elusive.
6. Every OrgZoo character has a place in the ecosystem, but they can become misaligned with context and desired outcomes.
7. Behaviors are not absolute (right or wrong): they can be perceived differently.
8. Culture is a subjective interpretation that usually has exceptions.
9. Behavioral DNA can be reliably visualized through inclusive cocreation activities.
10. Diversity of behavior can be (if facilitated well) a source of learning rather than conflict.
11. Unfamiliar behaviors can be learned and refined through role-play and gamified activities.
12. "Which character am I?" is a misinformed question — rather, ask "Which is best to be here and now?"

The principles cover many of the contexts in which you can use the OrgZoo creative metaphor to design activities that help people to expand their behavioral comfort zones. The list of principles is not intended to be comprehensive.

For further detail, see Shelley, A. (2022). *Becoming Adaptable: Creative facilitation to develop yourself and transform cultures.* Intelligent Answers.

Resource B: Neuroscience Findings Foundational to *Unleashing*

Neuroscience findings in the ICALS study are organized into 13 areas: Aging, Anticipating the Future, Creativity, Emotions, Epigenetics, Exercise and Health, Memory, Mirror Neurons, Plasticity, Social Support, Social Interaction, Stress, and the Unconscious. For purposes of this listing, when ICALS findings fit into two categories, they are listed in both of those categories. The numbers at the end of each finding refers to the page numbers in the *Unleashing* book where each finding is discussed. While they underly discussions in the Quest Field Guide, the neuroscience findings are not specifically pointed out.

Aging
At any age, mental exercise has a global positive effect on the brain. | 146
The best mental exercise to slow aging is new learning and doing things you've never done before. | 161
Despite certain cognitive losses, the engaged, mature brain can make effective decisions at more intuitive levels. | vii, 100, 161
Physical exercise reduces cognitive decline and dementia in older people. | 160

Anticipating the Future
The mind/brain creates an internal representation of the world. | 3
The neocortex constantly tries to predict the next experience. | 38, 229
Memory stores invariant forms (used to predict the future). | 130
The mind/brain unconsciously tailors internal knowledge to the situation at hand. | 20
The mind uses past learning and memories to complete incoming information. | 130

Creativity
An enriched environment can produce a personal internal reflective world of imagination and creativity. | 161
Extraordinary creativity can be developed. | 108
Conscious and unconscious patterns are involved with creativity. | 106, 111
Volleying between the conscious and the unconscious increases creativity. | 106, 111
The unconscious plays a big role in creativity. | 105
The unconscious produces flashes of insight. | 106, 212
Meditation quiets the mind. | 103, 217
Free-flow and randomly mixing patterns create new patterns. | 102, 199
Accidental associations can create new patterns. | 102

Emotions
The entire body is involved in emotions and the body drives the emotions. | 10, 165
Emotions can increase or decrease neuronal activity. | 171, 229, 270, 293
Unconscious interpretation of a situation can influence the emotional experience. | 173
Emotions influence all incoming information. | 9, 166, 270
Emotional tags impact memory. | 173
Emotional tags influence memory recall. | 9, 173
Emotions miss details but are sensitive to meaning. 174
Emotional fear inhibits learning. | 172, 229, 271
The brain can generate molecules of emotion to reinforce what is learned. | 183, 187

Epigenetics
What we believe leads to what we think leads to our knowledge, which leads to our actions, which determines outcomes. | 115, 229
Genes are not destiny. The environment can change the actions of genes via non-expression. | 114
Genes are operating options modulated by inputs from the environment, resulting in behavior. | 114

Exercise and Health
Exercise increases brainpower. | 146
Volition is necessary for benefit. Forced exercise does not promote neurogenesis. | 158
Meditation and other mental exercises can change feelings, attitudes and mindsets. | 103
Meditation quiets the mind. | 103, 217
Positive and negative beliefs affect every aspect of life. | 117
Physical and mental exercise and social bonding are significant sources of stimulation of the brain. | 193
Physical activity increases the number (and health) of neurons. | 159
Less than seven hours of sleep may impair memory. | 135

Memory
Memory is scattered throughout the entire cortex; it is not stored locally. | 129
Working memory is limited. | 127
Emotional tags impact memory. | 173

The brain stores only a part of the meaningful incoming sensory information. The gaps are filled in (re-created) when the memory is recalled. | 130
Memory stores invariant forms (used to predict the future). | 130
Memories are re-created each time they are recalled and therefore never the same. | 129
Memory recall is improved through temporal sequences of associated patterns, that is, stories and songs. | 132, 273
Repetition increases memory recall. | 146
Emotional tags influence memory recall. | 9, 173
Memory patterns cannot be erased at will. | 133
Memory patterns decay slowly with time. | 130
Less than seven hours of sleep may impair memory | 135

Mirror Neurons
Neurons create the same pattern when we see some action being taken as when we do it. | 195
Mirror neurons are a form of cognitive mimicry that transfers active behavior and other cultural norms. | 27, 196
What we see we become ready to do. | 194
Mirror neurons facilitate neural resonance between observed actions and executing actions. | 196
Rapid transfer of tacit knowledge (bypasses cognition). | 196, 215
Through [mental] reliving we recreate the feelings, perspectives and other phenomena that we observe. | 197
We may understand other people's behavior by mentally simulating it. | 197

Plasticity
Plasticity is a result of the connections between neural patterns in the mind and the physical world—what we think and believe impacts our physical bodies. | 7
Thoughts change the structure of the brain, and brain structure influences the creation of new thoughts. | 88
Learning depends on modification of the brain's chemistry and architecture. | 7
Maximum learning occurs when there are moderate levels of arousal—thereby initiating neural plasticity. | 149
An enriched environment increases the formation and survival of new neurons. | 161

Social Support
Language and social relationships build and shape the brain. | 201
Adults developing complex neural patterns need emotional support to offset discomfort of this process. | 192
The brain actually needs to seek out an affectively attuned other for learning. | 191
Affective attunement contributes to the evolution and sculpting of the brain. | 192
An enriched environment increases the formation and survival of new neurons. | 161

Social Interaction
Physical mechanisms have developed in our brain to enable us to learn through social interactions. | 189, 279
Social interaction mechanisms foster the engagement in affective attunement, consider the intentions of others, understand what another person is thinking and think about how we want to interact. | 191
Physical and mental exercise and social bonding are significant sources of stimulation of the brain. | 193

Stress
Stress depends on how we perceive a situation. | 150
Stress is an active monitoring system that constantly compares current events to past experiences. | 148
There is an optimum level of stress for learning (the inverted "U"). | 148
Stress focuses attention. | 147
Emotional fear inhibits learning. | 172, 229, 271
Voluntary learning promotes Theta waves that correlate with little or no stress and positive feedback. | 84
Belief systems can reduce stress through reducing uncertainty. | 117

The Unconscious
The unconscious brain is always processing. | 111, 122, 145, 216
Thinking is mostly unconscious. | 86, 216
A model of the self comes mostly from the unconscious. | 121
The unconscious never lies. | 125, 308
We may act for reasons we are not aware of. | 123
The unconscious can influence our thoughts and emotions without our awareness. | 123, 176
Unconscious interpretation of a situation can influence the emotional experience. | 173
The unconscious plays a big role in creativity. | 105
Conscious and unconscious patterns are involved with creativity. | 106, 111
The unconscious produces flashes of insight. | 106, 212
Volleying between the conscious and the unconscious increases creativity. | 106, 111

Resource C: Brief Field Guide Glossary

This brief glossary was recommended by OrgZoo critters to highlight terms they found particularly important during their Unleashing the Human Mind Workshop at MQI and on their ascent up Mountain Quest II:

Fundamental Words and Concepts:
(Terms discussed in the Introduction as Useful Definitions on pages vii to viii)
Brain
Consciousness
Knowledge
Learning
Mind
Mind/Brain
Mountains (as a metaphor)

Additional Words and Concepts (see Guide pages for information):
Attention and Intention | 31-32
Associative Patterning | iv
Complex Adaptive Systems | 3, 14
Consilience | 15
Consciousness Levels |37
Consciousness Shift | 12
Cortical Columns | 3,
Critical Thinking | 22
CUCA | 6
Deep Knowledge | 8
Heart-Mind Entrainment | 35
Hemispheric Synchronization | 21, 25
Humility | 51
ICALS | 5, 15-16, 25
Modes of Learning | 15
Mirror Neurons | 16, 41
Plasticity | 3
Quantum | 4, 11, 46, 62-63
Reference Frames | 3, 8, 58
Self | viii, 2, 24 (and throughout)
Self Phones | iv, 58
Soul | 29, 46-49
Spiritual | 3-4, 6, 47
Symbiotic Thinking | 51
Transreal Numbers | 50

Please note that a basic ICALS Glossary is available in *Expanding the Self: The Intelligent Complex Learning System by David Bennet, Alex Bennet and Robert Turner* (2015)

About the Authors

Dr. Alex Bennet is Professor on the faculty of Bangkok University's Institute for Knowledge and Innovation Southeast Asia (IKI-SEA) and Director of the Mountain Quest Institute. She is internationally recognized as an expert in knowledge management, change, and human and organizational systems. She served as Chief Knowledge Officer for the U.S. Department of the Navy, Chaired the Federal KM Working Group, and is recipient of the Distinguished Public Service Award. A Delta Epsilon Sigma and Golden Key National Honor Society graduate with diverse degrees, she believes in the multidimensionality and interconnectedness of humanity as we move into full consciousness. Favorite OrgZoo critter is Quercus robur, a continuous reminder of how much there is to learn. She may be contacted at alex@mountainquestinstitute.com

Robert Turner graduated magna cum laude from the Univ of Maryland and completed his Ed.M. at Boston Univ. He is a Phi Kappa Phi member—motto *"Let the love of learning rule humanity."* In the Army he founded the Army Fusion Center. At the FAA, he founded the Team Technology Center. His leadership vision grew out of memberships at The Institute for the Future and the IBM Institute for Knowledge Management. He chaired the Federal Knowledge Management Network and received the first national KM award. He is an MQI associate and avid genealogy consultant. Favorite OrgZoo critter is the Eagle because the Eagle has such a boundless range of vision. He may be contacted at turnerrg@hotmail.com

Dr. Arthur Shelley is a collaborative community builder, multi-awarded learning facilitator and creative education designer with over 30 years of professional experience across the international corporate, government and tertiary education sectors. He is the author of 4 books, has worked in 12 countries, is a mentor in several international communities, and has supervised PhD candidates in 5 countries. He has collaborated with organisations as diverse as NASA, Cirque Du Soleil, Local and National Governments, Universities, start-ups, SMEs, and multinational corporations. https://www.linkedin.com/in/arthurshelley/ Favorite OrgZoo critter: This is like asking a parent who is their favorite child! If only one, I feel that Owl is who I would identify with the most—knowledge sharing, mentoring and supportive of learning for others, as he displays throughout this text—behavior I have invested in most in my professional work. He may be contacted at Arthur@IntelligentAnswers.com.au

Jane Turner, proud mother of six, with her family now of 38, lived in Munich, Helsinki, and Kuala Lumpur for 10 years. She attended the Univ of Maryland and is a graduate of the Army Management Staff College (AMSC) graduate level Sustaining Base Leadership & Management Program. Her Civil Service career of 30 years culminated as an Operations Officer at AMSC. Community service included president of both her church's children's and women's organizations at the multi-congregation level. She served in election work for 30 years. She is an MQI associate, an editor, and avid genealogy consultant. Favorite OrgZoo critter is the Dog because Dog has unwavering loyalty.

Dr. Mark Boyes holds an Executive MBA (Distinction) and a PhD in Innovation. He has over 20 years of leadership experience in diverse project teams, information technology, business transformation and innovation. Mark has also been a Lecturer in Technology, Innovation Strategy and Design Thinking at RMIT's Graduate School of Business and Law. Mark has co-developed several of the world's first technologies and more recently is leading multi-million-dollar and award-winning digital health projects. Favorite OrgZoo critter is the snail, who is often hidden, quietly observing the ongoing activities.

By Arthur Shelley

The Organizational Zoo. A survival guide to workplace behaviour. 2007; 2021, 2nd ed. Intelligent Answers, Melbourne, Australia.

The Organizational Zoo concept is a set of creative metaphor characters that help develop understanding of human interactions and relationships. The "OrgZoo" characters represent behaviours and are used to facilitate constructive conversations about social interactions and the value of diversity. Activities described enable exploration of behavioural awareness and how our behavioural choices assist to engage more productively with each other. The Organizational Zoo book and Character Cards are most often used as a support resource for personal and professional development and for mentoring, coaching and self-development. It is a foundational companion book for Becoming Adaptable, a later book that adds more advanced activities and context around the application of the concepts.

Becoming Adaptable. Creative facilitation to develop yourself and transform cultures. 2021. Intelligent Answers, Melbourne, Australia.

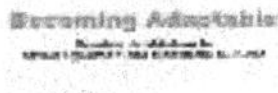

Becoming Adaptable provides insights on why adaptability enables better performance in personal and professional life. It details many insightful and practical facilitation activities to develop deeper understanding of people, relationships and culture. It explains how behavioural impacts what people perceive in context and what this means for how they interact with each other in learning and social interactions. The book explores how facilitation works best in practice, and what this means for the performance of individuals, teams and organisations. The new concept of MindFLEX is introduced to describe the parallel states of mind that enable a facilitator to more richly engages participants in facilitated activities. Becoming Adaptable is an essential resource for anyone wishing to make a difference for themselves and their peers, in both personal and professional situations. It is a compelling resource for experienced professionals who want to lead, coach, mentor or positively influence others around them.

KNOWledge SUCCESSion. Sustained performance and capability growth through strategic knowledge projects. 2017. Business Expert Press, USA.

KNOWledge SUCCESSion is a guide for executives and developing professionals who face the challenges of delivering business benefits for today, while building the capabilities required for an increasingly changing future. The book is structured to build from foundational requirements toward connecting the highly interdependent aspects of success in an emerging complex world. A wide range of concepts are brought together in a logical framework to enable readers of different disciplines to understand how they either create barriers or can be harvested to generate synergistic opportunities. The insights are robust as and pragmatic to help leaders create an environment in which their teams develop the knowledge and capabilities for sustained strategic success. This book also has extended learning for postgraduate students of business and project management in either an informal or a formal learning context. All successful medium to large organizations now need to have active management of projects and the ability to develop knowledge and capability to drive innovation and maintain relevance. There are detailed books on how to manage projects, texts of knowledge management, and volumes on innovation and change, but there is no one book that brings all these interdependent aspects of success together within the context of projects.

ALSO: ***Being a Successful Knowledge Leader. What knowledge practitioners need to know to make a difference.*** 2009. Ark Publishing, London, UK.

By David Bennet, Alex Bennet, Robert Turner

Unleashing the Human Mind: A Consilience Approach to Managing Self. 2022. MQIPress, Frost, West Virginia.

What does it mean to be human? Increasingly, we recognize that we are infinitely complex beings with immense emotional and spiritual, physical and mental capacities. Presiding over these human systems, our brain is a fully integrated, biological, and extraordinary organ that is preeminent in the known Universe. Its time has come. This book is grounded in the Intelligent Complex Adaptive Learning System (ICALS) theory based on over a decade of researching experiential learning through the expanding lens of neuroscience.

Review Bites from Around the World:

[Excerpted from the Foreword] *This volume is not a textbook of psychology or neuroscience, although it is based on their most recent discoveries, it is an active and contributory expression of a 'vivant'/living pulse throughout the progressive and interactive chapters, in a logical and sentient living 'heart'-like rhythm ... It is the mastery of the authors of this book to open the reader's mind and soul, thus offering the opportunity for the content of this live transmission to be discovered and interpreted in the most appropriate way by each reader ... so that only the reader's desire is needed to let him/her self be seduced by this wealth of wisdom, generously placed at the reader's disposal.*
-Dr. Florin Gaiseanu, Research Professor, Science and Technology of Information Bucharest (Romania) and Barcelona (Spain), Hoor Member of *NeuroQuanology* (Europe) and *International Journal of Neuropsychology and Behavioral Sciences* (USA)

With the backdrop of current global societal turmoil, the authors capture the urgency for changing course, and the research that provides a new path forward, which is found in the love of learning.
-John Lewis, Ed.D., CKO, Explanation Age LLC; Author of *Story Thinking*, USA

Every now and then a book comes along that compels your spirit. In these times of uncertainty and even great danger for humanity, this book reminds us of what it means to be human, our infinite potential and innate ability to learn and to love.
-Dr. Milton deSousa, Associate Professor, Nova School of Business and Economics, Portugal

This publication is the brilliant sublimination of a life-long accumulation of knowledge about the potential of our brain to learn, adapt and evolve to the best version of ourselves.
-Johan Cools, Higher Architecture Institute of Saint-Lucas Ghent, Belgium

In our time of fast societal, technological and environmental changes and disruption, experiential learning becomes one of the few solutions to quickly adapt and survive ... Such a rich and insightful book that will make you discover individual learning in a completely new way.
-Dr. Vincent Ribière, Managing Director of the Institute for Knowledge and Innovation Southeast Asia (IKI-SEA), Bangkok University, Thailand

In Unleashing, David Bennet, Alex Bennet and Robert turner expand our collective understanding of the Human Mind by embedding timeless wisdoms, transliterations, and their collective expansive knowledge and experiences that enrich our lives from cover to cover.
-Bob Beringer, CEO of EOR, Intelligence Professional, Author of *Linux Clustering with CSM and GPFS*, USA

Once in a while, I am exposed to a work so profound that it literally causes a massive shift in my own thinking and beliefs ... I found myself riveted to each paragraph as I embarked on a journey that vastly deepened my understanding of the learning process.
-Duane Nickull, Author, Technologist, and Seeker of Higher Truth, Canada

The existential growth risks of the Anthropocene call for a reinvention of learning as the primary means of advancing human development. It could not be timelier for this book ... the path towards knowledge-based consilience appears to be both urgent and potentially endless.
-Dr. Francisco Javier Carrillo Gamboa, President, World Capital Institute, Brazil.

Very few people have the gift to integrate such complex ideas, especially those about learning ... this work can be likened to the Webb Telescope, which gives us more clarity into our mysteries. Well worth the viewing!
-Michael Stankosky, DSc, Author, Philosopher, Professor, Editor-Emeritus, Member of the Academy of Scholars, USA

Other books from MQI Press

The Profundity and Bifurcation of Change by Alex Bennet, David Bennet, Arthur Shelley, Theresa Bullard and John Lewis.

Part I: Laying the Groundwork
Part II: Learning from the Past
Part III: Learning in the Present
Part IV: Co-Creating the future
Part V: Living the Future (with Donna Panucci)

This series supports progression of the Intelligent Social Change Journey (ISCJ), a developmental journey of the body, mind and heart, moving from the heaviness of cause-and-effect linear extrapolation to the fluidity of co-evolving with our environment, to the lightness of breathing our thought and feelings into reality. Grounded in development of our mental faculties, these are phase changes, each building and expanding pr3vious learning in our movement toward intelligent activity.

The little Conscious Look Books to the right are specifically focused on sharing key concepts from The Profundity series and looking at what those concepts mean to YOU!

MQIPress Conscious Look Books:

The Possibilities that are YOU!

by Alex Bennet

All Things in Balance
The Art of Thought Adjusting
Associative Patterning and Attracting
Beyond Action
Connections as Patterns
Conscious Compassion
The Creative Leap
The Emerging Self
The Emoting Guidance System
Engaging Forces
The ERC's of Intuition
Grounding
The Humanness of Humility
Intention and Attention
Knowing
The Living Virtues of Today
Me as Co-Creator
Seeking Wisdom
Staying on the Path
Transcendent Beauty
Truth in Context

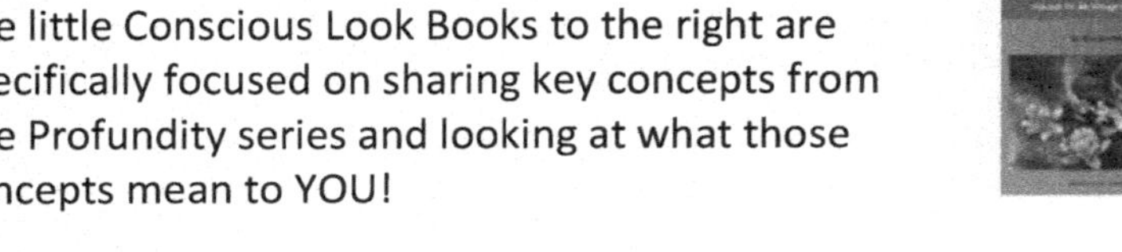

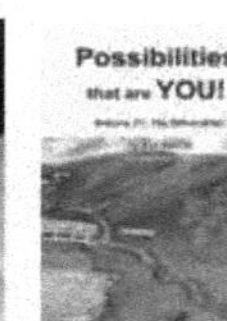

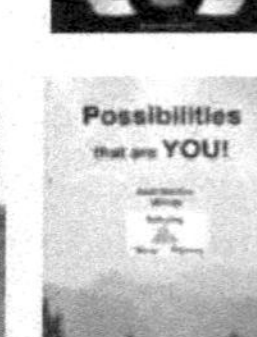

The QUEST: Where the Mountains meet the Library by the Drs. David and Alex Bennet (with Afterword by Robert Turner). 2021. MQIPress, Frost, WV., USA.

So here it is. This is a book of big ideas, the very ideas that have continuously filled our minds and hearts over the past 20 years, bubbling up and down as we traveled the world, then settling into printed text when we returned home to Mountain Quest where the mountains meet the library, a beautiful valley set in the Allegheny Mountains of West Virginia.

Each chapter is, it seems, a chapter of our lives, yet all of it blends together in a magnificent unfolding of thought and learning. How to share 20 years in nine chapters? Impossible, I think. Yet, YOU have been through those 20 years as well, on the same experiential roller coaster that has both titillated and challenged our survival. A global roller coaster still raging today through the highest points and most perilous curves.

ALSO AVAILABLE through MQI Press:

The Course of Knowledge: A 21st Century Theory (2015)
Decision-Making in The New Reality: Complexity, Knowledge and Knowing (2013)
Expanding the Self: The Intelligent Complex Adaptive Learning System (2015)
Leading with the Future in Mind: Knowledge and Emergent Leadership (2015)
With Passion, We Live and Love: Research, Prose, Verse and Music (2021)

The Quest to elevate our minds, to unleash the human brain, is as lofty as the mountains.

Made in USA - North Chelmsford, MA
1350320_9781949829655
12.28.2022 1245